Algebra For Dummies®

Cheat Sheet

Relationship Symbols

= Equal

≠ Not equal

≅ Approximately

< Less than

≤ Less than or equal to

> Greater than

≥ Greater than or equal to

Order of Operations

1. Powers or roots
2. Multiplication or division
3. Addition or subtraction

Graphing Formulas

Midpoint $M = \dfrac{x_1 + x_2}{2}, \dfrac{y_1 + y_2}{2}$

Distance $d = \sqrt{(x_2 - x_1)^2 + (y_2 - y_1)^2}$

Slope $m = \dfrac{y_2 - y_1}{x_2 - x_1}$

Rules of Exponents

$$x^a \cdot x^b = x^{a+b}$$

$$x^a \div x^b = x^{a-b}$$

$$x^0 = 1$$

$$x^{-1} = \tfrac{1}{x}$$

$$(x^a)^b = x^{ab}$$

$$\sqrt[b]{x^a} = x^{a/b}$$

Special Factoring Rules

$$x^2 - y^2 = (x+y)(x-y)$$

$$x^3 - y^3 = (x-y)(x^2+xy+y^2)$$

$$x^3 + y^3 = (x+y)(x^2-xy+y^2)$$

$$x^2 + 2xy + y^2 = (x+y)2$$

Special Formulas

Quadratic Formula $x = \dfrac{-b \pm \sqrt{b^2 - 4ac}}{2a}$

Pythagorean Theorem $a^2 + b^2 = c^2$

First 100 Prime Numbers

2	11	53	101	151	211	251	307	353	401	457	503
3	13	59	103	157	223	257	311	359	409	461	509
5	17	61	107	163	227	263	313	367	419	463	521
7	19	67	109	167	229	269	317	373	421	467	523
	23	71	113	173	233	271	331	379	431	479	541
	29	73	127	179	239	277	337	383	433	487	
	31	79	131	181	241	281	347	389	439	491	
	37	83	137	191		283	349	397	443	499	
	41	89	139	193		293			449		
	43	97	149	197							
	47			199							

Algebra For Dummies®

Cheat Sheet

First 20 Perfect Squares

$1^2 = 1$	$11^2 = 121$
$2^2 = 4$	$12^2 = 144$
$3^2 = 9$	$13^2 = 169$
$4^2 = 16$	$14^2 = 196$
$5^2 = 25$	$15^2 = 225$
$6^2 = 36$	$16^2 = 256$
$7^2 = 49$	$17^2 = 289$
$8^2 = 64$	$18^2 = 324$
$9^2 = 81$	$19^2 = 361$
$10^2 = 100$	$20^2 = 400$

First 10 Perfect Cubes

$1^3 = 1$
$2^3 = 8$
$3^3 = 27$
$4^3 = 64$
$5^3 = 125$
$6^3 = 216$
$7^3 = 343$
$8^3 = 512$
$9^3 = 729$
$10^3 = 1,000$

Everyday Formulas

Formula	Translation
$A = \pi r^2$	Area of a circle equals π times the radius squared
$C = \pi d$	Circumference of a circle equals π times the diameter
$A = lw$	Area of a rectangle equals length times width
$D = rt$	Distance traveled equals the rate multiplied by the time
$I = prt$	Simple interest earned equals principal times rate times time
$A = P\left(1 + \frac{r}{n}\right)^{nt}$	The amount resulting from compounding interest equals the Principal times the sum of 1 and the quotient of the rate of interest divided by the number of times compounded each year all raised to the product of the number of times compounded times the term (number of years)
$F = (\%)C + 32$	The temperature in degrees Fahrenheit equals ⅗ times the degrees Celsius plus 32
$P = R - C$	Profit equals revenue minus cost
$R = xp$	Revenue equals the number of items sold, x, times the price of each item
$S = dp$	Savings equals discount times price
$C = p + tp$	Total cost equals price plus tax percentage times price

For Dummies: Bestselling Book Series for Beginners

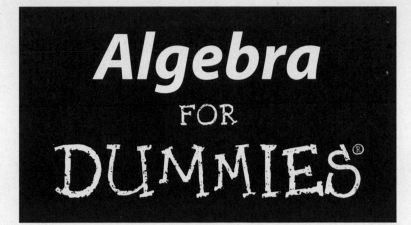

Algebra FOR DUMMIES®

by Mary Jane Sterling

WILEY

Wiley Publishing, Inc.

Algebra For Dummies®

Published by
Wiley Publishing, Inc.
111 River Street
Hoboken, NJ 07030
www.wiley.com

Copyright © 2001 by Wiley Publishing, Inc., Indianapolis, Indiana

Published by Wiley Publishing, Inc., Indianapolis, Indiana

Published simultaneously in Canada

For general information on our other products and services or to obtain technical support, please contact our Customer Care Department within the U.S. at 800-762-2974, outside the U.S. at 317-572-3993, or fax 317-572-4002.

Wiley also publishes its books in a variety of electronic formats. Some content that appears in print may not be available in electronic books.

Library of Congress Cataloging-in-Publication Data:

Library of Congress Control Number: 2001089362

ISBN: 0-7645-5325-9

Manufactured in the United States of America

15 14 13 12 11 10 9 8 7

1O/QR/RQ/QT/IN

About the Author

Mary Jane Sterling has been an educator since graduating from college. Teaching at the junior high, high school, and college levels, she has had the full span of experiences and opportunities while working in education. She has been teaching at Bradley University in Peoria, Illinois, for the past twenty years.

Dedication

The author would like to dedicate this book to her husband, Ted, and three children Jon, Jim, and Jane, for their demonstrations of love and support. She also dedicates it to two teachers, Catherine Kay and Alba Biagini, who are responsible for the professional path she's taken. And, finally, the book is dedicated to her nephew, Timothy, for his demonstrations of courage and faith.

Author's Acknowledgments

The author would like to thank several people for making this book possible: Mike Kantor for his thorough technical review and many folks at Hungry Minds: Roxane Cerda for starting this whole thing, Susan Decker for launching the project, and especially Kathleen Dobie and Esmeralda St. Clair for their pulling, pushing, tugging, and Herculean feats of editing to get the project in shape.

Publisher's Acknowledgments

We're proud of this book; please send us your comments through our Online Registration Form located at www.dummies.com/register.

Some of the people who helped bring this book to market include the following:

Acquisitions, Editorial, and Media Development

Project Editor: Kathleen A. Dobie

Acquisitions Editors: Roxane Cerda, Susan L. Decker

Copy Editor: Esmeralda St. Clair

Technical Editor: Mike Kantor

Editorial Manager: Christine Meloy Beck

Editorial Assistant: Jennifer Young

Cover Photo: ©PictureQuest

Production

Project Coordinator: Nancee Reeves

Layout and Graphics: Joyce Haughey, Jill Piscitelli, Betty Schulte, Rashell Smith, Erin Zeltner

Proofreaders: Andy Hollandbeck, Betty Kish, Susan Moritz, Carl Pierce, Dwight Ramsey

Indexer: Liz Cunningham

Publishing and Editorial for Consumer Dummies

Diane Graves Steele, Vice President and Publisher, Consumer Dummies
Joyce Pepple, Acquisitions Director, Consumer Dummies
Kristin A. Cocks, Product Development Director, Consumer Dummies
Michael Spring, Vice President and Publisher, Travel
Brice Gosnell, Associate Publisher, Travel
Suzanne Jannetta, Editorial Director, Travel

Publishing for Technology Dummies

Richard Swadley, Vice President and Executive Group Publisher
Andy Cummings, Vice President and Publisher

Composition Services

Gerry Fahey, Vice President of Production Services
Debbie Stailey, Director of Composition Services

Contents at a Glance

Cartoons at a Glance

By Rich Tennant

page 153

page 7

page 321

page 87

page 241

Cartoon Information:
Fax: 978-546-7747
E-Mail: richtennant@the5thwave.com
World Wide Web: www.the5thwave.com

Table of Contents

Introduction

*T*ell me the truth. When you got up this morning, did you really expect to be reading this introduction to an algebra book? Was it high on your list of things to do? I'm glad you're reading this, but why?

You're probably in one of two situations:

- ✔ You've taken the plunge and bought the book.
- ✔ You're checking things out before committing to the purchase.

In either case, you'd probably like to have some good, concrete reasons why you should go to the trouble of reading and finding out about algebra.

One of the most commonly asked questions in a mathematics classroom is, "What will I ever use this for?" Some teachers can give a good, convincing answer. Others hem and haw and stare at the floor. My favorite answer is, "Algebra gives you *power*." Algebra gives you the *power* to move on to bigger and better things in mathematics. Algebra gives you the *power* of knowing that you know something that your neighbor doesn't know. Algebra gives you the *power* to be able to help someone else with an algebra task or to explain to your child about these logical mathematical processes. Algebra is a system of symbols and rules that is universally understood, no matter what the spoken language. Algebra provides a clear, methodical process that can be followed from beginning to end. It's an organizational tool that is most useful when followed with the appropriate rules. What *power!*

This book isn't like a mystery novel; you don't have to read it from beginning to end. In fact, you can peek at how it ends and not spoil the rest of the story. This book is divided into some general topics from the beginning nuts and bolts to the important tool of factoring to equations and applications. I've tried to use many examples, each a bit different from the others, and each showing a different *twist* to the topic. The examples have explanations to aid your understanding.

The vocabulary I use is mathematically correct and understandable. Along with the *how-to,* I hope you'll find the *why.* Sometimes it's easier to recall a process if you understand why it works and don't just try to memorize a meaningless list of steps.

About This Book

If you're looking for help with some of the basic tools of algebra, you can find that type of information in the first part of the book. Think of these tools as being like what a cook needs. You can't cook a soufflé unless you know how to whisk the eggs and turn on the oven. Your success later depends on your preparation. Of course, you may be beyond these basics. Great! How about the second part?

In the second part, I spend a lot of time explaining factoring. Factoring is really no more than changing what the expression looks like. And the factored form is one where everything is all multiplied together. You can find which of the factoring techniques you need to brush up on if you get stuck with a problem.

And where are the equations, you may ask. In Part III, I give you any type of equation you desire, in order from simplest to most complex. More rules and methods are added as the equations get more difficult. I also throw in inequalities for good measure.

Part IV covers a good deal of the answer to the question, "What is this algebra stuff used for?" The applications in these chapters are more on the practical side — discussing things you may actually experience.

The Part of Tens serves as a nice set of lists of game plans. You may need only one thing in the list, or you may run down through the whole thing, in order. Use them as you wish.

Have fun with this. Think of this book as being like a computer's "Help" button. If you have a problem, you can find the answer (hopefully, better explained than some of those computer "helps").

Following the conventions

You find two ways of expressing numbers or numerals: In the descriptions, numbers ten and under and math operators (plus sign, equals signs, and so on) are spelled out. In problems and examples, though, I use the actual digits and symbols. This should make for an easier read.

Terms special to algebra are italicized and defined. You also find these terms in the glossary at the back of the book, for easy reference.

I offer a lot of step-by-step instruction to make things really clear. I often set out general steps, then offer a couple of examples so that you get a sense of how to use the steps in different situations.

Eyeing those icons

The silly little drawings in the margin of the book are there to draw your attention to specific text. The icons I use in this book are:

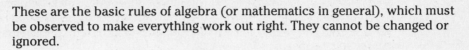

These are the basic rules of algebra (or mathematics in general), which must be observed to make everything work out right. They cannot be changed or ignored.

These paragraphs, often attached to sidebars, offer facts you may find interesting, but which you do not need to know. These tidbits are absolutely not important, but they make algebra a little less impersonal and ethereal.

These paragraphs help clarify a symbol or process. I may discuss the topic in another section of the book, or just be reminding you of a basic algebra rule. If you haven't been in school for a while, you may find that the name of a process has changed.

This indicates a definition or clarification for a step in a process, a technical term, or an expression.

The info next to this icon isn't life-or-death important, but it generally can help make your life easier — at least your life in algebra.

This alerts you to something that can be particularly tricky. Errors crop up frequently when working with the process or topic next to this icon, so I call special attention to it with this icon.

What Not to Read

You can get a lot from this book by just going from icon to icon. The lines next to the Algebra Rules icon pull everything together very tidily. It's when you want more detail that you want to read between the icons.

The sidebars (those little gray boxes) include some historical stuff — mathematicians' lives may not make for exciting movies, but some mathematicians have done some rather interesting things. There are also some of my favorite anecdotes and stories to lighten the mood. You can pick and choose rather easily because of the general format.

Foolish Assumptions

I'm not going to assume that you're as crazy about math as I am — you may be even *more* excited about it! I do assume, though, that you have a mission here — to brush up on your skills, improve your mind, or just have some fun. I also assume that you have some experience with algebra — full exposure for a year or so, maybe a class you took a long time ago, or even just some preliminary concepts.

If you went to high school in the United States, you probably took an algebra class. If you're like me, you can distinctly remember your first (or only) algebra teacher. This is a momentous time in your life. I can remember Miss McDonald saying, "This is an *n*." My whole secure world of numbers was suddenly turned upside-down. You may be delving into the world of algebra again to refresh those long-ago lessons. Is your son or daughter coming home with assignments that are beyond your memory? Never fear. Help is here!

How This Book Is Organized

Where do you find what you need quickly and easily? This book is divided into sections dealing with the most frequently discussed and studied concepts of basic algebra.

Part 1: Starting Off with the Basics

The "founding fathers" of algebra based all their rules on the assumption that everyone would *agree* on some things first. In language, for instance, we all agree that the English word for *good* means the same thing whenever it appears. The same goes for algebra. Everyone uses the same rules of addition, subtraction, multiplication, division, fractions, exponents, and so on. The algebra wouldn't work if the basic rules were different for different people. We wouldn't be able to communicate. The basics in this part are to help you review what all these things are that everyone has agreed on over the years.

This is where you find the basics of arithmetic, fractions, powers, and signed numbers. These are the tools necessary to be able to deal with the material that comes later. The review of basics here puts a spin on the more frequently used algebra techniques. You can refer to these chapters when working through material in the later chapters.

In these first chapters, I introduce you to the world of letters and symbols. Studying the use of the symbols and numbers is like studying a new language. There's a vocabulary, some frequently used phrases, and some cultural applications. The language is the launching pad for further study.

Part II: Figuring Out Factoring

Part II contains the factoring and simplifying. Algebra has few processes more important than factoring. This is a way of rewriting expressions to help make solving the problem easier. Factoring is where expressions are changed from addition and subtraction to multiplication and division. The easiest way to solve many problems is to work with the wonderful multiplication property of zero. It basically says that to get a zero you multiply by zero. Seems simple, and yet it's really grand.

There are simple factorings where you just have to recognize a similarity. There are more complicated factorings where you not only have to recognize a pattern, but have to know the rule to use. But, don't worry, I fill you in on all the differences.

Part III: Working Equations

This is where you get into the nitty-gritty of finding answers. Some methods for solving equations are elegant; some are down-and-dirty. I show you many types of equations and many methods for solving them. Usually I give you one method for solving each type of equation, but I present alternatives when it makes sense also, so that you can see that some ways are better than others.

Part IV: Applying Algebra

The whole point of doing algebra is in this part. There are everyday formulas — and not-so-everyday formulas. There are familiar situations — and situations that may be totally unfamiliar. I don't have space to show you every possible type of problem, but I give you enough practical uses, patterns, and skills to prepare you for almost any situation you encounter.

Part V: The Part of Tens

Here I give you the ten most common errors made in algebra, the ten ways to factor a quadratic equation, ten of the most commonly used rules of divisibility, and ten steps to solving a story problem. These are neat checklists that you can use as a quick reference. For details, you look in the middle of the book.

I also throw in a glossary of algebra and math terms in the Appendix.

Where to Go from Here

If you want to refresh your basic skills or boost your confidence, then start with Part I. If you're ready for some factoring practice and need to pinpoint which method to use with what, then go to Part II. Part III is for those who are ready to solve equations; you can find just about any type you're ready to attack. Part IV is where the good stuff is — applications — things to do with all those good solutions. The lists in Part V are usually what you'd look at after visiting one of the other parts, but why not start here? It's a fun place!

Studying algebra can give you some logical exercises. As you get older, the more you exercise your brain cells, the more alert and "with it" you remain. "Use it or lose it" means a lot in terms of the brain. What a good place to use it, right here!

The best *why* for studying algebra is just that it's beautiful. Yes, you read that right. Algebra is poetry, deep meaning, and artistic expression. Just look, and you'll find it. Also, don't forget that it gives you *power*.

Welcome to algebra! Enjoy the adventure!

Part I
Starting Off with the Basics

The 5th Wave By Rich Tennant

In this part . . .

How many of you can just up and go on a trip to a foreign country on a moment's notice? Not many, I suspect. It takes preparation. You need to get your passport renewed, apply for a visa, pack your bags, and arrange for someone to take care of your cats. It takes preparation for the trip to turn out well.

The same is true of algebra: It takes preparation for the experience to turn out well. Careful preparation prevents problems along the way. In this part, you find the essentials you need to have a successful trip.

Chapter 1

Assembling Your Tools

You probably have heard the word *algebra* on many occasions and knew that it had something to do with mathematics. Perhaps you remember that algebra has enough information to require taking two separate high school algebra classes — Algebra I and Algebra II. But what exactly *is* algebra? What is it *really* used for?

This chapter answers these questions and more, providing the straight scoop on some of the contributions to algebra's development, what it's good for, how algebra is used, and what tools you need to make it happen.

In a nutshell, *algebra* is a way of generalizing arithmetic. Through the use of variables that can generally represent *any* value in a given formula, general formulas can be applied to *all* numbers. Algebra uses positive and negative numbers, integers, fractions, operations, and symbols to analyze the relationships between values. It's a systematic study of numbers and their relationship, and it uses specific rules.

For example, the formula $a \times 0 = 0$ shows that any real number, represented here by the *a*, multiplied by zero always equals zero. (For more information on the multiplication property of zero, see Chapter 14.)

In algebra, by using an *x* to represent the number two, for example in $x + x + x = 6$, you can generalize with the formula $3x = 6$.

You may be thinking, "That's great and all, but come on. Is it really *necessary* to do that — to plop in letters in place of numbers and stuff?" Well, yes. Early mathematicians found that using letters to represent quantities simplified problems. In fact, that's what algebra is all about — simplifying problems.

The basic purpose of algebra has been the same for thousands of years: to allow people to solve problems with unknown answers.

Aha algebra

Dating back to about 2000 B.C. with the Babylonians, algebra seems to have developed in slightly different ways in different cultures. The Babylonians were solving three-term quadratic equations, while the Egyptians were more concerned with linear equations. The Hindus made further advances in about the sixth century A.D. In the seventh century, Brahmagupta of India provided general solutions to quadratic equations and had interesting takes on zero. The Hindus regarded irrational numbers as actual numbers — although everybody didn't hold to that belief at that time.

The sophisticated communication technology that exists in the world now was not available then, but early civilizations still managed to exchange information over the centuries. In A.D. 825, al-Khowarizmi of Bagdad wrote the first algebra textbook. One of the first solutions to an algebra problem, however, is on an Egyptian papyrus that is about 3,500 years old. Known as the Rhind Papyrus after the Scotsman who purchased the 1-foot-wide, 18-foot-long papyrus in Egypt in 1858, the artifact is preserved in the British Museum — with a piece of it in the Brooklyn Museum. Scholars determined that in 1650 B.C., the Egyptian scribe Ahmes copied some earlier mathematical works onto the Rhind papyrus.

One of the problems reads, "Aha, its whole, its seventh, it makes 19." The "aha" isn't an exclamation. The word "aha" designated the unknown. Can you solve this early Egyptian problem? It would be translated, using current algebra symbols, as: $x + \frac{x}{7} = 19$. The unknown is represented by the x, and the solution is $x = 16\frac{5}{8}$. It's not hard; it's just messy.

Beginning with the Basics: Numbers

Where would mathematics and algebra be without numbers? A part of everyday life, numbers are the basic building blocks of algebra. Numbers give you a value to work with.

Where would civilization be today if not for numbers? Without numbers to figure the total cubits, Noah couldn't have built his ark. Without numbers to figure the distances, slants, heights, and directions, the pyramids would never have been built. Without numbers to figure out navigational points, the Vikings would never have left Scandinavia. Without numbers to examine distance in space, humankind could not have landed on the moon.

Even the simple tasks and the most common of circumstances require a knowledge of numbers. Suppose that you wanted to figure the amount of gasoline it takes to get from home to work and back each day. You need a number for the total miles between your home and business and another number for the total miles your car can run on one gallon of gasoline.

The different sets of numbers are important because what they look like and how they behave can set the scene for particular situations or help to solve particular problems. It's sometimes really convenient to declare, "I'm only going to look at whole-number answers," because whole numbers do not include fractions. This may happen if you're working through a problem that involves a number of cars. Who wants half a car?

Algebra uses different sets of numbers, such as whole numbers and those that follow here, to solve different problems.

Really real numbers

Real numbers are just what the name implies. In contrast to imaginary numbers, they represent *real* values — no pretend or make-believe. Real numbers, the most inclusive set of numbers, comprise the full spectrum of numbers; they cover the gamut and can take on any form — fractions or whole numbers, decimal points or no decimal points. The full range of real numbers includes decimals that can go on forever and ever without end. The variations on the theme are endless.

For the purposes of this book, I always refer to real numbers.

Counting on natural numbers

A *natural* number is a number that comes naturally. What numbers did you first use? Remember someone asking, "How old are you?" You proudly held up four fingers and said, "Four!" The natural numbers are also *counting* numbers: 1, 2, 3, 4, 5, 6, 7, and so on into infinity.

You use natural numbers to count items. Sometimes the task is to count how many people there are. A half-person won't be considered (and it's a rather grisly thought). You use natural numbers to make lists.

Wholly whole numbers

Whole numbers aren't a whole lot different from the natural numbers. The *whole* numbers are just all the natural numbers plus a zero: 0, 1, 2, 3, 4, 5, and so on into infinity.

Whole numbers act like natural numbers and are used when whole amounts (no fractions) are required. Zero can also indicate none. Algebraic problems often require you to round the answer to the nearest whole number. This makes perfect sense when the problem involves people, cars, animals, houses, or anything that shouldn't be cut into pieces.

Integrating integers

Integers allow you to broaden your horizons a bit. Integers incorporate all the qualities of whole numbers and their opposites, or additive inverses of the whole numbers (refer to the "Operating with opposites" section in this chapter for information on additive inverses). *Integers* can be described as being positive and negative whole numbers: $\ldots -3, -2, -1, 0, 1, 2, 3, \ldots$.

Integers are popular in algebra. When you solve a long, complicated problem and come up with an integer, you can be joyous because your answer is probably right. After all, it's not a fraction! This doesn't mean that answers in algebra can't be fractions or decimals. It's just that most textbooks and reference books try to stick with nice answers to increase the comfort level and avoid confusion. This is the plan in this book, too. After all, who wants a messy answer, even though, in real life, that's more often the case.

Being reasonable: Rational numbers

Rational numbers act rationally! What does that mean? In this case, acting rationally means that the decimal equivalent of the rational number behaves. The decimal ends somewhere, or it has a repeating pattern to it. That's what constitutes "behaving." Some rational numbers have decimals that end in 2, 3.4, 5.77623, −4.5. Other rational numbers have decimals that repeat the same pattern, such as $3.164164164\ldots = 3.\overline{164}$, or $.666666666\ldots = .\overline{6}$. The horizontal bar over the 164 and the 6 lets you know that these numbers repeat forever.

In *all* cases, rational numbers can be written as a fraction. They all have a fraction that they are equal to. So one definition of a *rational number* is any number that can be written as a fraction.

Restraining irrational numbers

Irrational numbers are just what you may expect from their name — the opposite of rational numbers. An *irrational* number *cannot* be written as a fraction, and decimal values for irrationals never end and never have a nice pattern to them. Whew! Talk about irrational! For example, pi, with its never-ending decimal places, is irrational.

Evening out even and odd numbers

An *even* number is one that divides evenly by two. "Two, four, six, eight. Who do we appreciate?"

Digits, fingers, and toes through history

The Hindu-Arabic numerals, such as 1, 2, 3, 4, 5, 6, 7, 8, 9, originated with the Hindus and were created to go along with a decimal system. The word *decimal* comes from the Latin word meaning *tenth* or *tithe*. The Hindu-Arabic system is a *positional* system, which means that the order you write digits in matters. The number 35 is different from the number 53 because in 35, the 3 stands for three tens and in 53, the three stands for three ones.

The main reason humans developed a decimal, or base-ten, system is because humans usually have ten fingers and ten toes. It could have been a base-twenty system or a base-five system — like the Babylonians had. From about 1700 B.C. until about A.D. 500, however, most scientists used a base-sixty system. Using sixty as a base came about because the number of

days in a year was estimated to be roughly 360 days, and sixty was one of the nice divisors of 360. Remnants of the early base-sixty system are found in our minutes and seconds. Can you imagine having to remember sixty different digits instead of just ten?

The symbols used in the early systems all stood for something: one of these, two of those, and so on. For a long while, there was no digit or symbol for nothing or zero. The first symbol for zero (something like an upside-down W) wasn't introduced until about 300 B.C. Before then, to indicate that there was nothing, the writer left an empty space, which wasn't very efficient. Sometimes the writer forgot to leave a space, and a careless writer may not have left enough of a space. Plus, there was no clear way to indicate more than one zero.

An *odd* number is one that *does not* divide evenly by two. The even and odd numbers alternate when you list all the integers.

Varying Variables

Variable is the most general word for a letter that represents the unknown, or what you're solving for in an algebra problem. A variable *always* represents a number.

Algebra uses letters, called *variables,* to represent numbers that correspond to specific values. Usually, if you see letters toward the beginning of the alphabet in a problem, such as *a, b,* or *c,* they represent known or set values, and the letters toward the end of the alphabet, such as *x, y,* or *z,* represent the unknowns, things that can change, or what you're solving for.

The following list goes through some of the more commonly used variables.

✔ An *n* doesn't really fall at the beginning or end of the alphabet, but it's used frequently in algebra, often representing some unknown quantity or number — probably because *n* is the first letter in *number*.

✔ The letter *x* is often the variable you solve for, maybe because it's a letter of mystery: X marks the spot, the x-factor, *The X Files*. Whatever the reason *x* is so popular as a variable, the letter also is used to indicate multiplication. You have to be clear, when you use an *x*, that it isn't taken to mean multiply.

✔ *C* and *k* are two of the more popular letters used for representing known amounts or constants. The letters that represent variables and numbers are usually small case: *a, b, c,* and so on. Capitalized letters are used most commonly to represent the answer in a formula, such as the capital *A* for area of a circle equals pi times the radius squared, $A = \pi r^2$. (You can find more information on the area of a circle in Chapter 17.) The letter *C,* mentioned previously as being a popular choice for a constant, is used frequently in calculus and physics, and it's capitalized there — probably more due to tradition than any good reason.

Speaking in Algebra

Algebra and symbols in algebra are like a foreign language. They all mean something and can be translated back and forth as needed. It's important to know the vocabulary in a foreign language; it's just as important in algebra.

✔ An *expression* is any combination of values and operations that can be used to show how things belong together and compare to one another. $2x^2 + 4x$ is an example of an expression.

✔ A *term,* such as 4*xy,* is a grouping together of one or more factors (variables and/or numbers). Multiplication is the only thing connecting the number with the variables. Addition and subtraction, on the other hand, separate terms from one another. For example, the expression $3xy + 5x - 6$ has three *terms.*

✔ An *equation* uses a sign to show a relationship — that two things are equal. By using an equation, tough problems can be reduced to easier problems and simpler answers. An example of an equation is $2x^2 + 4x = 7$. See the chapters in Part III for more information on equations.

✔ An *operation* is an action performed upon one or two numbers to produce a resulting number. Operations are addition, subtraction, multiplication, division, square roots, and so on. See Chapter 6 for more on operations.

✔ A *variable* is a letter that always represents a number, but it varies until it's written in an equation or inequality. (An *inequality* is a comparison of two values. See more on inequalities in Chapter 16.) Then the fate of the variable is set — it can be solved for, and its value becomes the solution of the equation.

✔ A *constant* is a value or number that never changes in an equation — it's constantly the same. Five (5) is a constant because it is what it is. A variable can be a constant if it is assigned a definite value. Usually, a variable representing a constant is one of the first letters in the alphabet. In the equation $ax^2 + bx + c = 0$, a, b, and c are constants and the x is the variable. The value of x depends on what a, b, and c are assigned to be.

✔ An *exponent* is a small number written slightly above and to the right of a variable or number, such as the 2 in the expression 3^2. It's used to show repeated multiplication. An exponent is also called the *power* of the value. For more on exponents, see Chapter 4.

Taking Aim at Algebra Operations

In algebra today, a variable represents the unknown (see more on variables in the "Speaking in Algebra" section earlier in this chapter). Before the use of symbols caught on, problems were written out in long, wordy expressions. Actually, using signs and operations was a huge breakthrough. First, a few operations were used, and then algebra became fully symbolic. Nowadays, you may see some words alongside the operations to explain and help you understand, like having subtitles in a movie. Look at this example to see what I mean. Which would you rather write out:

The number of quarts of water multiplied by six and then that value added to three

or

$6x + 3$?

I'd go for the second option. Wouldn't you?

By doing what early mathematicians did — letting a variable represent a value, then throwing in some operations (addition, subtraction, multiplication, and division), and then using some specific rules that have been established over the years — you have a solid, organized system for simplifying, solving, comparing, or confirming an equation. That's what algebra is all about: That's what algebra's good for.

> # What's in a word?
>
> The word *algebra* is a variation on the word *al-jabr*, an Arabic word, which roughly means a reunion or joining together of parts. This word was changed further when the Moors brought the word *algebrista*, meaning *bonesetter* (someone who joins or puts together bones), to Spain in the the Middle Ages. Signs over barber shops in Spain saying *Algebrista y Sangradoe* indicated that the shop offered a bonesetter *and* bloodletter. At that time, and for centuries, barbers performed minor medical procedures to supplement their income. The traditional red-and-white-striped barber pole symbolized blood and bandages. Maybe that's why the word *algebra*, which comes from *algebrista*, has a reputation for being painful at times.

Deciphering the symbols

The basics of algebra involve symbols. Algebra uses symbols for quantities, operations, relations, or grouping. The symbols are shorthand and are much more efficient than writing out the words or meanings. But you need to know what the symbols represent, and the following list shares some of that info.

- ✔ + means *add* or *find the sum, more than,* or *increased by;* the result of addition is the *sum.*

- ✔ − means *subtract* or *minus* or *decreased by* or *less;* the result is the *difference.*

- ✔ × means *multiply* or *times.* The values being multiplied together are the *multipliers* or *factors;* the result is the *product.* Some other symbols meaning multiply can be grouping symbols: ( ), [ ], { } ·, *. In algebra, the × symbol is used infrequently because it can be confused with the variable *x*. The dot is popular because it's easy to write. The grouping symbols are used when you need to contain many terms or a messy expression. By themselves, the grouping symbols don't mean to multiply, but if you put a value in front of a grouping symbol, it means to multiply. For more on the grouping symbols, skip ahead to the "Grouping" section.

- ✔ ÷ means *divide.* The number that's going into the *dividend* is the *divisor.* The result is the *quotient.* Other signs that indicate division are the fraction line and slash, /.

- ✔ $\sqrt{}$ means to take the *square root* of something — to find the number, which multiplied by itself gives you the number under the sign (see more on square roots in Chapter 4).

- ✔ | | means to find the *absolute value* of a number, which is the number itself or its distance from zero on the number line (see more on *absolute value* in Chapter 2).

✔ . . . means *et cetera, and so on,* or *in the same pattern.* You use an ellipsis in algebra when you have a long list of numbers and don't want to have to write all of them. For instance, if you want to list numbers starting with 1 and going up by 1 forever and ever, write: "1, 2, 3, 4, . . . ". Or you can write the list of numbers from 600 through 1,000 as, "600, 601, 602, . . . , 1,000."

✔ π is the Greek letter pi that refers to the irrational number: 3.14159 . . . . It represents the relationship between the diameter and circumference of a circle. For more information on this relationship, see Chapter 17.

Grouping

When a car manufacturer puts together a car, several different things have to be done first. The engine experts have to construct the engine with all its parts. The body of the car has to be mounted onto the chassis and secured, too. Other car specialists have to perform the tasks that they specialize in as well. When these tasks are all accomplished in order, then the car can be put together. The same thing is true in algebra. You have to do what's inside the *grouping* symbol before you can use the result in the rest of the equation.

Grouping symbols tell you that you have to deal with the *terms* inside the grouping symbols *before* you deal with the larger problem.

The main grouping symbols are

✔ ( ) Parenthesis (This one is used most often.)

✔ [ ] Brackets

✔ { } Braces

For example, $8 - (4 - 2)$ says to do what's in the parenthesis first. This is different from $(8 - 4) - 2$. The first expression works out to be 6, and the second expression to 2.

These three grouping symbols — the parenthesis, bracket, and brace — are used both alone and with each other. When used together, the symbols organize a more complicated problem.

Defining relationships

Algebra is all about relationships — not the he-loves-me-he-loves-me-not kind of relationship — but the relationships between numbers or among the terms of an equation. Although algebraic relationships can be just as complicated as romantic ones, you have a better chance of understanding an algebraic relationship. The symbols for the relationships are given here.

The beginning of the equal sign

Robert Recorde first used the equal sign (=) in the mid-1500s. He wrote, "I will sette as I doe often in woorke use, a paire of parrallels, or Gemowe lines of one lengthe, thus ==, because noe 2 thynges can be moare equalle." However, not all mathematicians immediately accepted the equal sign. Some preferred two upright parallel lines. A symbol resembling α (with longer "tails") was also popular for quite a while. The equal sign seemed to have been generally accepted in the mid-1600s.

✔ = means that the first value *is equal to* or the same as the value that follows.

✔ ≠ means that the first value *is not equal to* the value that follows.

✔ ≈ means that one value is *approximately the same* or *about the same* as the value that follows; this is used when rounding numbers.

✔ ≤ means that the first value is *less than or equal to* the value that follows.

✔ < means that the first value is *less than* the value that follows.

✔ ≥ means that the first value is *greater than or equal to* the value that follows.

✔ > means that the first value is *greater than* the value that follows.

Operating with opposites

When solving equations in algebra, doing the *opposite* to work your way toward the answer comes up often. You have to *undo* operations that have been *done* to the variable. The opposite of an operation is another operation that gets you back where you started. This is used primarily to get rid of numbers that are combined with a variable so you can solve for the variable in an equation.

Being contrary: Doing opposite operations

The opposite of adding three is subtracting three. If you add three to 100, you get 103. If you then subtract three from 103, you're back where you started.

✔ The opposite of addition is subtraction.

✔ The opposite of subtraction is addition.

✔ The opposite of multiplication is division.

✔ The opposite of division is multiplication.

  ✔ The opposite of taking a square root is *squaring* (multiplying a value by itself).

  ✔ The opposite of squaring is taking the square root.

  ✔ The opposite of cubing is taking the cube root.

Dealing with the opposites of numbers

A number actually has two *opposites:* the *additive inverse* and the *multiplicative inverse:*

  ✔ The *additive inverse* is the number with the opposite sign. So −3 is the additive inverse of 3, and 16 is the additive inverse of −16. Use these if 3 or 16 is being added to a variable, and you want to get the variable alone; this is used when solving an equation for the value of the variable.

  ✔ The *multiplicative inverse* is also called the reciprocal. The *reciprocal* is the original number written as the bottom of a fraction with a one on the top. So $\frac{1}{2}$ is the reciprocal of 2, and 25 is the reciprocal of $\frac{1}{25}$. If a number starts out as a fraction, its reciprocal is just that number written upside-down. So the reciprocal of $\frac{4}{7}$ is $\frac{7}{4}$. Use this if a number multiplies or divides a variable; it gets the variable alone so it can be solved for.

Playing by the Rules

The basics of algebra also involve rules, like the rules to follow when you're driving. If everyone follows the same rules, accidents and chaos are less likely. The same goes for algebra. You have to observe the rules of algebra when you work with variables, numbers, and symbols. Following the rules is especially important when you solve problems because you don't know what number a variable stands for. The rules were developed, and everyone uses the same ones as everyone else, which is why the language of algebra is so universal.

Algebra involves symbols, such as variables and operation signs, which are the tools that you can use to make algebraic expressions more useable and readable. These things go hand in hand with simplifying, factoring, and solving problems, which are easier to solve if broken down into basic parts. Using symbols is actually much easier than wading through a bunch of words.

  ✔ To *simplify* means to combine all that can be combined, cut down on the number of terms, and put an expression in an easily understandable form. To find more on simplifying, see Chapter 13.

  ✔ To *factor* means to change two or more terms to just one term. See Part II for more on factoring.

  ✔ *Solve* means to find the answer. In algebra, it means to figure out what the variable stands for.

Equation solving is fun because there's a point to it. You solve for something (often a variable, such as x) and get an answer that you can check to see whether you're right or wrong. It's like a puzzle. It's enough for some people to say, "Give me an x." What more could you want? But solving these equations is just a means to an end. The real beauty of algebra shines when you solve some problem in real life — a practical application. Are you ready for these two words: *story problems?* Story problems are the whole point of doing algebra. Why do algebra unless there's a good reason? Oh, I'm sorry. Some of you may just like to solve algebra equations for the fun alone. Yes, there are folks like that. But some folks love to see the way a complicated paragraph in the English language can be turned into a neat, concise expression, such as, "The answer is three bananas."

Going through each step and using each tool to play this game is entirely possible. *Simplify, factor, solve, check.* That's good! Lucky you. It's time to dig in!

Chapter 2

Assigning Signs:
Positive and Negative Numbers

*N*umbers have many characteristics: They can be big, little, even, odd, whole, fraction, positive, negative, and sometimes cold and indifferent — I'm kidding about that last one. Chapter 1 describes numbers' different names and categories. But this chapter concentrates on just the positive and negative characteristics.

Positive and *negative* are words you use and hear every day:

> "You have a *positive* influence on me."

> "I'm getting *negative* vibes."

This chapter tells you how to add, subtract, multiply, and divide signed numbers, no matter whether all the numbers are all the same sign or mix-and-matched.

Showing Some Signs

Early on, mathematicians realized that using plus and minus signs and making rules for their use was going to be a big advantage in their number world. They also realized that if they used the minus sign, there was no need to create a bunch of completely new symbols for negative numbers. After all, positive and negative numbers are related to one another, and the slight addition of the minus sign works well. Negative numbers have positive counterparts and vice versa. This means that –3 and +3 are related. A new symbol, such as a ϑ, didn't have to be created to represent the opposite of three — they could just use the minus sign. If you have a handle on what having six bananas would be like, then you can imagine what *not* having six bananas would be like, also.

Numbers that are opposite in sign but the same otherwise are *additive inverses*.

Two numbers are *additive inverses* of one another if their sum is zero, $a + (-a) = 0$. Additive inverses are always the same distance from zero (in opposite directions) on the number line. For example, the additive inverse of –6 is +6; the additive inverse of $+\frac{1}{5}$ is $-\frac{1}{5}$.

Picking out positive numbers

Positive numbers are bigger, greater, or higher than zero. They are on the opposite side of zero from the negative numbers. If you arrange a tug-of-war between positive and negative numbers, the positive numbers line up on the right side of zero, as Figure 2-1 shows.

Figure 2-1:
Some
positive
numbers all
lined up.

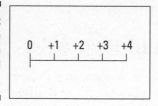

Positive numbers get bigger and bigger the farther they are from zero: 81 is bigger than 25 because it's farther away from zero; 212° F, the boiling temperature of water, is farther away from zero than 32° F, the temperature at which water freezes. They're both positive numbers, but one may seem more positive than the other. Check out the difference between freezing water and boiling water to see how much more positive a number can be!

Making the most of negative numbers

The concept of a number less than zero can be difficult to grasp. Sure you can say "less than zero," and even write a book with that title, but what does it really mean? Think of entering the ground floor of a large government building. You go to the elevator and have to choose between going up to the first, second, third, or fourth floors, or going down to the first, second, third, fourth, or fifth subbasement (down where all the secret stuff is). The farther you are from the ground floor, the farther the number of that floor is from zero. The second subbasement could be called floor –2, but that may not be a good number for a floor.

Negative numbers are smaller than zero. On a line with zero in the middle, negative numbers line up on the left, as shown in Figure 2-2.

Figure 2-2:
Negative numbers lining up on the left.

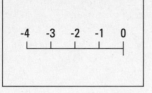

Negative numbers get smaller and smaller the farther they are from zero. This can get confusing because you may think that –400 is *bigger* than –12. But just think of –400° F and –12° F. Neither is anything pleasant to think about, but –400° is definitely less pleasant — colder, lower, smaller.

Regarding negative numbers, the number closer to zero is the *bigger* number.

Comparing positives and negatives

Although my mom always told me not to compare myself to other people, comparing numbers to other numbers is often useful. And, when you compare numbers, the greater-than (>) and less-than (<) signs come in handy, which is why I use them in Table 2-1, where I put some positive and negative signed numbers in perspective.

Table 2-1	Comparing Positive and Negative Numbers
Comparison	*What It Means*
6 > 2	6 is greater than 2; 6 is farther from 0 than 2 is.
10 > 0	10 is greater than 0; 10 is positive and is bigger than 0.
–5 > –8	–5 is greater than –8; –5 is closer to 0 than –8 is.

(continued)

Table 2-1 *(continued)*

Comparison	What It Means
−300 > −400	−300 is larger than −400.
0 > −6	−6 is negative and is smaller than 0.
7 > −80	Positive numbers are always bigger than negative numbers.

So, putting the numbers 6, −2, −18, 3, 16, and −11 in order from smallest to biggest gives you: −18, −11, −2, 3, 6, and 16, which are shown as dots on a number line in Figure 2-3.

Figure 2-3:
Positive and negative numbers on a number line.

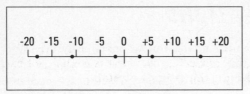

Zeroing in on zero

But what about zero? I keep comparing numbers to see how far they are from zero. Is zero positive or negative? The answer is that it's neither. Zero has the unique distinction of being neither positive nor negative. Zero separates the positive numbers from the negative ones — what a job!

FUN FACT

Plus (+) and minus (−) signs

The first time plus (+) and minus (−) signs appeared in print, they referred to surpluses and deficits in business situations — not to arithmetic operations. But even before these signs appeared formally in print, the plus and minus symbols were used extensively. For example, plus and minus signs were painted on barrels of goods to indicate whether the barrels were full or not.

Negatives

Early Chinese civilizations were responsible for many important inventions and discoveries. The Chinese are credited with developing gun powder, printing, paper, and the compass. They also had a good way to deal with negative numbers — long before negative numbers were formally recognized.

The Chinese used two sets of calculating rods — a red set for positive coefficients and a black set for negative coefficients. For some reason, the colors were eventually reversed, and now red indicates a financial deficit or negative (such as "in the red") and black means to the good or positive.

Going In for Operations

Operations in algebra are nothing like operations in hospitals. Well, you get to dissect things in both, but dissecting numbers is a whole lot easier (and a lot less messy) than dissecting things in a hospital.

Algebra is just a way of generalizing arithmetic, so the operations and rules used in arithmetic work the same for algebra. Some new operations crop up, though, to make things more interesting than just adding, subtracting, multiplying, and dividing. I'm going to introduce one of those new operations after explaining the difference between a binary operation and a nonbinary operation.

Breaking in to binary operations

Relax, I haven't suddenly switched to *Astronomy For Dummies.* The *binary* in this section refers to operations with two numbers, not systems with two stars.

Bi means two. A *bi*cycle has two wheels. A *bi*gamist has two spouses. A *bi*nary operation involves two numbers. Addition, subtraction, multiplication, and division are all *binary operations* because you need two numbers to perform them. You can add 3 + 4, but you can't add 3 + if there's nothing after the plus sign. You need another number.

Introducing nonbinary operations

A *nonbinary* operation needs just one number. A nonbinary operation performs a task and spits out the answer. Square roots are nonbinary operations. You find $\sqrt{4}$ by performing this operation on just one number. See Chapter 4 for more on square roots.

Making your own binary operation

All it takes to create a binary operation is to make up a rule as to how it works and what numbers can be used. For example, you could say, "I have a new binary operation named star, *. When you star two numbers together, you put a 0 between them." For example, 4*7 = 407. As you can see 4*7 is not the same as 7*4.

Now, you may say, what good is this operation? None that I can see. Maybe this gives you more of an appreciation for the binary operations that already exist. They do something useful.

One of the most important nonbinary operations is finding the absolute value of a number. The *absolute value* operation tells you how far a number is from zero. It doesn't pay any attention to whether the number is less than or greater than zero; it just determines *how far* it is from zero.

The symbol for absolute value is two vertical bars: | |.

The absolute value of *a*, where *a* represents any real number, either positive or negative, is

- $|a| = a$, where $a \geq 0$
- $|a| = -a$, where $a < 0$ (negative), and $-a$ is positive.

Doing absolute value operations looks like this:

- $|3| = 3$
- $|-4| = 4$
- $|-87| = 87$
- $|0| = 0$

Basically, the absolute value operation tells you how far the number is from zero. It doesn't pay any attention to whether the number is less than zero or greater than zero; it just determines *how far* from zero.

Operating with Signed Numbers

If you're on an elevator in a building with four floors above the ground floor and five floors below ground level, you can have a grand time riding the elevator all day, pushing buttons, and actually "operating" with signed numbers.

You're probably too young to remember this, but people actually used to get paid to ride elevators and push buttons all day. I wonder if these people had to understand algebra first. If you want to go up five floors from the third sub-basement, you end up on the second floor above ground level.

Adding like to like: Same-signed numbers

When your first grade teacher taught you that one plus one equals two, she probably didn't tell you that this was just one part of the whole big addition story. She didn't mention that adding one positive number to another positive number is really a special case. If she *had* told you this big story stuff — that you can add positive and negative numbers together or add any combination of positive and negative numbers together — you might have packed up your little school bag and sack lunch and left the room right then and there.

Adding all positive numbers is just a small part of the whole addition story, but it was enough to get you started at that time. This section gives you the big story — all the information you need to add signed numbers.

The first thing to consider in adding signed numbers is to start with the easiest situation: The numbers have the same sign. Look at what happens:

- ✔ You have 3 apples and your friend gives you 4 apples.

  $(+3) + (+4) = +7$

  You now have 7 apples.
- ✔ You owed Jon $8 and had to borrow $2 more.

  $(-8) + (-2) = -10$

  Now you're $10 in debt.

There's a nice "S" rule for this addition. See if you can say it quickly three times in a row: When the signs are the same, you find the sum, and the sign of the sum is the same as the signs.

This rule holds when *a* and *b* represent any two real numbers:

$$(+a) + (+b) = +(a + b)$$
$$(-a) + (-b) = -(a + b)$$

I wish I had something as alliterative for all the rules, but this is math — not poetry!

Say you're adding –3 and –2. The signs are the same; so you find the sum of 3 and 2, which is 5. The sign of this sum is the same as the signs of –3 and –2, so the *sum* is also a negative.

Check out these examples:

- $(+8) + (+11) = +19$. The signs are all positive.
- $(-14) + (-100) = -114$. The sign of the sum is the same as the signs.
- $(+4) + (+7) + (+2) = +13$. Because all the numbers are positive, add them and make the sum positive, also.
- $(-5) + (-2) + (-3) + (-1) = -11$. This time all the numbers are negative, so add them and give the sum a minus sign.

Adding same-signed numbers is a snap! (A little more alliteration for you.)

Adding different signs

Can a relationship between a Leo and a Gemini ever add up to anything? I don't know the answer to that question, but I do know that numbers with different signs add up very nicely. You just have to know how to do it, and in this section, I tell you.

When the signs of two numbers are different, forget the signs for a while and find the *difference* between the numbers. This is the difference between their *absolute values*. (For a refresher on absolute values, turn to the "Introducing nonbinary operations" section earlier in this chapter.) The number farther from zero determines the sign of the answer.

$(+a) + (-b) = +(|a| - |b|)$ if the positive a is farther from zero.

$(+a) + (-b) = -(|b| - |a|)$ if the negative b is farther from zero.

Look what happens when you add numbers with different signs:

- You had $20 in your wallet and spent $12 for your theatre ticket.

  $(+20) + (-12) = +8$

  After settling up, you have $8 left.
- I have $20, but it costs $32 to fill my car's gas tank.

  $(+20) + (-32) = -12$

  I'll have to borrow $12 to fill the tank.

The following examples give you some more combinations:

- $(+6) + (-7) = -1$. The difference between 6 and 7 is 1. Seven is farther from 0 than 6 is, so the answer is –1.
- $(-6) + (+7) = +1$. This time the 7 is positive. It's still farther from 0 than the 6. The answer this time is +1.

> ✔ $(-4) + (+3) + (+7) + (-5) = +1$. If you take these in order from left to right (although you can add in any order you like), you add the first two together to get -1. Add that to the next to get $+6$. Then add this to the last number to get $+1$.

Subtracting signed numbers

Subtracting signed numbers is really easy to do: You *don't!* Instead of inventing a new set of rules for subtracting signed numbers, mathematicians determined that it's easier to change the subtraction problems to addition problems and use the rules I explained in the previous section.

Think about that for a moment. Just change the subtraction problem into an addition problem. It doesn't make much sense, does it? Everybody knows that you can't just change an arithmetic operation and expect to get the same or right answer. You found out a long time ago that $10 - 4$ isn't the same as $10 + 4$. You can't just change the operation and expect it to work out correctly.

So, to make this work you really change *two* things to even things out.

When subtracting signed numbers, change the minus sign to a plus sign *and* change the number that the minus sign was in front of to its opposite. Then just add the numbers using the rules for adding signed numbers. See Chapter 1 for more on opposites.

> ✔ $(+a) - (+b) = (+a) + (-b)$
>
> ✔ $(+a) - (-b) = (+a) + (+b)$
>
> ✔ $(-a) - (+b) = (-a) + (-b)$
>
> ✔ $(-a) - (-b) = (-a) + (+b)$

Coming up with nothing

Consider adding two numbers with different signs where there is *no difference* between the absolute value of the numbers:

$$(+3) + (-3)$$

$$(-5) + (+5)$$

The difference between the numbers without their signs is zero. And because zero is neither positive nor negative — it has no sign — that takes care of having to determine what the sign

of the answer is by which is farther from zero. Neither wins! So, in the following examples, zero is the hero.

$$(-10) + (+10) = 0$$

$$(-a) + (+a) = 0$$

$$(+abc) + (-abc) = 0$$

In the last two examples, assume that a, b, and c are the same throughout the expression.

The following examples put these concepts into real-life terms:

✔ The submarine was sixty feet below the surface when the skipper shouted, "Dive!" It went down another 40 feet.

$-60 - (+40) = -60 + (-40) = -100$.

Change from subtraction to addition. Change the 40 to its opposite, -40. Then use the addition rule. The submarine is now 100 feet below the surface.

✔ Kids play a version of "Mother may I?" where players may ask, "Mother, may I take three steps forward?" A "Yes" answer allows the player to move three steps closer to Mother. A "No" answer means the player takes three steps backward. A player may ask, "Mother, may I take four steps backward?" In this case, a "No" answer means take four steps forward. The net result of these two answers is

$(-3) - (-4) = (-3) + (+4) = +1$

Change the -4 to its opposite to change from subtraction to addition. The player is one step closer to Mother after the two moves.

To subtract signed numbers, change the minus sign to a plus sign and change the sign of the number that follows.

Multiplying and dividing signed numbers

Multiplication and division are really the easiest operations to do with signed numbers. As long as you can multiply and divide, the rules are not only simple, but they're also the same for both operations.

When multiplying and dividing two signed numbers: if the two signs are the same, then the result is *positive;* when the two signs are different, then the result is *negative:*

$$(+a) \times (+b) = +ab \qquad (+a) \div (+b) = +(a \div b)$$
$$(+a) \times (-b) = -ab \qquad (+a) \div (-b) = -(a \div b)$$
$$(-a) \times (+b) = -ab \qquad (-a) \div (+b) = -(a \div b)$$
$$(-a) \times (-b) = +ab \qquad (-a) \div (-b) = +(a \div b)$$

Notice in which cases the answer is positive and in which cases it's negative. Also, notice that multiplication and division seem to be "as usual" except for the positive and negative signs. Check out the following examples:

✔ $(-8) \times (+2) = -16$

✔ $(-5) \times (-11) = +55$

✔ $(+24) \div (-3) = -8$

✔ $(-30) \div (-2) = +15$

You can mix up these operations doing several multiplications or divisions or a mixture of each and use the following Even-Odd Rule.

Even-Odd Rule: When multiplying and dividing a bunch of numbers, count the number of negatives to determine the final sign. An *even* number of negatives means the result is *positive*. An *odd* number of negatives means the result is *negative*. The following examples show you how it's done:

✔ $(+2) \times (-3) \times (+4) = -24$: This problem has just one negative sign. Because one is an odd number (and often the loneliest number), the answer is negative.

✔ $(+2) \times (-3) \times (+4) \times (-1) = +24$: Two negative signs mean a positive answer because two is an even number.

✔ $\dfrac{(+4) \times (-3)}{(-2)} = +6$: An even number of negatives means a positive answer.

✔ $\dfrac{(-12) \times (-6)}{(-4) \times (+3)} = -6$: Three negatives yield a negative.

✔ $(-1)(-1)(-1)(-1)(-1)(-1)(-1)(-1)(-1)(-1)(-1)(-1)(-1)(-1)(-1) = -1$: An odd number of negative signs gives you a negative answer.

Finally, you can prove to your mother that sometimes two wrongs *do* make a right!

Working with Nothing: Zero and Signed Numbers

What role does zero play in the signed number show? What does it do to the signs of the answers? Well, when you're doing addition or subtraction, what zero does depends on where it is. When you multiply or divide, zero tends to just wipe out the numbers and leave you with zero.

Some general guidelines about zero:

✔ **Adding zero:** $0 + a$ is just a. Zero doesn't change the value of a. (This is also true for $a + 0$.)

✔ **Subtracting zero:** $0 - a = -a$. Use the rule for subtracting signed numbers: Change the operation from subtraction to addition and change the sign of the second number. Likewise, $a - 0 = a$. It doesn't change the value of a to subtract zero from it.

✔ **Multiplying by zero:** $a \times 0 = 0$. If you're in a club with a bunch of friends and none of you has anything, multiplying what each of you has yields nothing: Likewise, $0 \times a = 0$.

Multiplying any number by zero always yields zero.

✔ **Dividing by zero:** $0 \div a = 0$. Take you and your friends: If none of you has anything, dividing that *nothing* into shares just means that each share has nothing. And you can't use zero as the divisor because if you have a things, you can't divide them into zero parts.

So, working with zero isn't too tricky. You follow normal addition and subtraction rules, and just keep in mind that multiplying and dividing with zero leaves you with nothing — literally.

Associating and Commuting with Expressions

Algebra operations follow certain rules, and those rules have certain properties. In this section, I talk about two of those properties — the commutative property and the associative property.

Reordering operations: The commutative property

Before discussing the commutative property, take a look at the word *commute.* You probably commute to work or school and know that whether you're going from home to work or from work to home, the distance is the same: The distance doesn't change because you change directions (although getting home during rush hour may make that distance *seem* longer).

The same principle is true of *some* algebraic operations: It doesn't matter whether you add $1 + 2$ or $2 + 1$; the answer is still 3. Likewise, multiplying 2×3 or 3×2 yields 6.

The *commutative property* means that you can change the order of the numbers in an operation without affecting the result. Addition and multiplication are commutative. Subtraction and division are not. So,

$$a + b = b + a$$

$$a \times b = b \times a$$

$$a - b \neq b - a \text{ (except in a few special cases)}$$

$$a \div b \neq b \div a \text{ (except in a few special cases)}$$

In general, subtraction and division are *not* commutative. The special cases occur when you choose the numbers carefully. For instance, if *a* and *b* are the same number, then the subtraction appears to be commutative because switching the order doesn't change the answer. In the case of division, if *a* and *b* are opposites, then you get –1 no matter which order you divide them in. This is why, in mathematics, big deals are made about proofs. A few special cases of something may work, but a real rule or theorem has to work all the time.

Look at the following examples:

✔ 4 + 5 = 9 and 5 + 4 = 9 so 4 + 5 = 5 + 4

✔ $3 \times (-7) = -21$ and $(-7) \times 3 = -21$ so $3 \times (-7) = (-7) \times 3$

✔ $(-5) - (+2) = (-7)$ and $(+2) - (-5) = +7$ so $(-5) - (+2) \neq (+2) - (-5)$

✔ $(-6) \div (+1) = -6$ and $(+1) \div (-6) = -1/6$ so $(-6) \div (+1) \neq (+1) \div (-6)$

Keep in mind that the commutative property holds true only for addition and multiplication.

Associating expressions: The associative property

The commutative property has to do with the order of the numbers when you perform an operation. The associative property has to do with how the numbers are grouped when you perform an operation on more than two numbers.

Think about what the word *associate* means. When you associate with someone, you're close to the person, or you form a group with the person. Say that Anika, Becky, and Cora associate. Whether Anika drives over to pick up Becky and the two of them go to Cora's and pick her up, or Cora is at Becky's house and Anika picks up both of them at the same time, the same result occurs — the same people are in the car at the end.

The *associative property* means that if the grouping of the operation changes, the result remains the same. (If you need a reminder about grouping, check out Chapter 1.) Addition and multiplication are associative. Subtraction and division are *not* associative operations. So,

$$a + (b + c) = (a + b) + c$$

$$a \times (b \times c) = (a \times b) \times c$$

$$a - (b - c) \neq (a - b) - c \text{ (except in a few special cases)}$$

$$a \div (b \div c) \neq (a \div b) \div c \text{ (except in a few special cases)}$$

You can always find a few cases where the property works even though it isn't supposed to. For instance, in the subtraction problem $5 - (4 - 0) = (5 - 4) - 0$ the property seems to work. Also, in the division problem $6 \div (3 \div 1) = (6 \div 3) \div 1$, it seems to work. Although there are exceptions, a rule must work *all* the time.

Some real-number examples may make this clearer:

- ✔ $4 + (5 + 8) = 4 + 13 = 17$ and $(4 + 5) + 8 = 9 + 8 = 17$
  So $4 + (5 + 8) = (4 + 5) + 8$

- ✔ $3 \times (2 \times 5) = 3 \times 10 = 30$ and $(3 \times 2) \times 5 = 6 \times 5 = 30$
  So $3 \times (2 \times 5) = (3 \times 2) \times 5$

- ✔ $13 - (8 - 2) = 13 - 6 = 7$ and $(13 - 8) - 2 = 5 - 2 = 3$
  So $13 - (8 - 2) \neq (13 - 8) - 2$

- ✔ $48 \div (16 \div 2) = 48 \div 8 = 6$ and $(48 \div 16) \div 2 = 3 \div 2 = \dfrac{3}{2}$
  So $48 \div (16 \div 2) \neq (48 \div 16) \div 2$

The commutative and associative properties come in handy when you work with algebraic expressions. You can change the order of some numbers or change the grouping to make the work less messy or more convenient. Just keep in mind that you can commute and associate addition and multiplication operations, but not subtraction or division.

Chapter 3

Figuring Out Fractions and Dealing with Decimals

At one time or another, most math students wish that the world was made up of whole numbers only. But those non-whole numbers called fractions really make the world a wonderful place. (Well, that may be stretching it a bit.) In any case, fractions are here to stay, and this chapter helps you delve into them in all their wondrous workings.

Compare developing an appreciation of fractions with watching or playing a sport: If you want to enjoy and appreciate a game, you have to understand the rules. You know that this is true if you watch soccer games. That off-sides rule is hard to understand. But finally you figure it out, discover the basics of the game, and love the sport. This chapter gets down to basics with the rules involving fractions so you can *play the game.* You may not think that decimals belong in a chapter on fractions, but there's no better place for them. Decimals are just shorthand for the most favorite fractions. Words that are often used are abbreviated, such as Mr., Dr., Tues., Oct., and so on! Likewise, fractions with denominators of 10, 100, 1,000, and so on are abbreviated with decimals.

Pulling Numbers Apart and Piecing Them Back Together

Understanding fractions, where they come from, and why they look the way they do helps when you're working with them. A fraction has two parts:

$$\frac{top}{bottom} \text{ or } \frac{numerator}{denominator}$$

The *denominator*, or bottom number, tells you the total number of items. The *numerator*, or top number, tells you how many of the total (the bottom number) are being considered.

Perhaps you can remember the exact placement of the numbers and their proper names if you think in terms of

- **N:** **N**umerator; **N**orth; ↑
- **D:** **D**enominator; **D**own; ↓

In all the cases using fractions, the denominator tells you how many *equal* portions or pieces there are. Without the *equal* rule, you could get different pieces in various sizes. For instance, in a recipe calling for $\frac{1}{2}$ cup of flour, if you didn't know that the one part was one of two *equal* parts, then there could be two *unequal* parts — one big and one little. Should the big or the little part go into the cookies?

Along with terminology like *numerator* and *denominator,* fractions fall into one of three types: *proper, improper,* and *mixed,* which are covered in the following sections.

Making your bow to proper fractions

The simplest type of fraction to picture is a proper fraction, which is always just part of one whole thing. One whole pie can be cut into proper fractions. One whole play can be divided into fractions — acts or scenes.

Roman fraction bars

The Romans didn't use the symbols used now to indicate fractions, such as the fraction line and numbers, but the Romans recognized fractions and had words for them:

$\frac{11}{12}$ was *deunx,* or $\frac{1}{12}$ taken away

$\frac{10}{12}$ was *dextans,* or $\frac{1}{6}$ taken away

$\frac{9}{12}$ was *dodrans,* or $\frac{1}{4}$ taken away

$\frac{8}{12}$ was *bi,* or $\frac{1}{3}$ taken away

It's interesting that the Romans perceived fractions as what's taken away — perhaps they were more inclined to see a glass half empty than half full.

Fibonacci, during the thirteenth century, was the first European mathematician to use the fraction bar as it's used today. The bar is found in manuscripts from the Middle Ages, but in printing it was often omitted — probably because of typographical difficulties.

In a *proper fraction*, the numerator is always smaller than the denominator, and its value is always less than one.

Take a look at the following proper fractions:

- $\frac{5}{6}$: Cut a cake into six slices (six shows how many things equal the total). Eat one piece, and you still have five pieces left. Lucky you! You can have your cake and eat it too!
- $\frac{4}{12}$: You took four months out of last year to finish the project.
- $\frac{1}{16}$: One pound of butter equals 16 ounces. Put one ounce of butter on the popcorn.

Remember: A minute on the lips, a lifetime on the hips!

Getting to know improper fractions

An *improper* fraction has more parts than necessary for one whole number, which has nothing to do with a lack of social decorum. These top-heavy fractions, however, are useful in many situations. The bottom number tells you what size the pieces are. It's just that in the case of improper fractions, there are more than enough pieces to make one whole number.

Improper fractions are fractions whose numerators, or tops, are bigger than their denominators.

- $\frac{15}{8}$: After the party, Maria put all the leftover pieces of pizza together. There were 15 pieces, each $\frac{1}{8}$ of a pizza.
- $\frac{4}{3}$: Doubling the amount of sugar in the recipe requires four measures of sugar, each $\frac{1}{3}$ of a cup.

They may be called improper, but these fractions behave very well.

Mixing it up with mixed numbers

Improper fractions can get a bit awkward. Mixed numbers help clean up the act. Using a *mixed* number — one with both a whole number and a fraction — to express the same thing that an improper fraction expresses makes things easier to deal with. For example, instead of using the improper fraction $\frac{4}{3}$, you can use the mixed number $1\frac{1}{3}$. Recipes are easier to use; hat sizes are easier to read.

Converting fractions on Wall Street

Stocks in the U.S. Stock Market used to be priced using fractions for the parts. You'd see prices, such as $16\frac{3}{4}$, and read that the price had gone down by $\frac{5}{8}$. This custom of using fractions is supposed to have started when coins could be broken into pieces; it's easier to break something in half, then the half into halves (quarters), then the quarters into halves (eighths), and so on.

In the year 2000, the stock market changed these parts to decimals. This wasn't any response to a desire to go metric. It was just to make the increments smaller. There are eight divisions between each number and the next if you use eighths and ten divisions using decimals, or tenths. The tenths are smaller than eighths, so there are smaller steps going up (or down).

A *mixed number* contains both a whole number *and* a fraction, as the following examples show.

- $4\frac{1}{2}$: The recipe calls for four and one-half cups of flour.
- $7\frac{3}{8}$: The hat size is 7 plus $\frac{3}{8}$ more so it isn't too tight.
- $5\frac{7}{12}$: It's been 5 years and 7 months since he left for Europe.

Following the Sterling Low-Fraction Diet

Just when you thought you'd heard about all the possible ways to reduce, here comes another. Reduce! If only weight reduction were this easy!

When you use fractions, you want them to be as nice as possible. In this case, *nice* means the smallest possible numbers in the numerator and denominator of the fraction. Sometimes small numbers are just easier to deal with — easier to understand and easier to visualize — than larger numbers. Doing the arithmetic is much easier with smaller numbers, too.

The lowest terms are desirable when borrowing money; the lowest terms are also desirable when dealing with fractions. A fraction is in *lowest terms* if no number (other than 1) divides both the numerator and denominator evenly.

Figuring out equivalent fractions

When you multiply or divide the numerator and denominator of a fraction by the same number, you don't change the value of the fraction. In fact, you're basically multiplying or dividing by one because any time the numerator and denominator of a fraction are the same number, it equals one. If you divide the numerator and denominator of $\frac{16}{32}$ by 4, you're basically dividing $\frac{16}{32}$ by $\frac{4}{4}$, which equals one.

- ✔ $\frac{4}{5}$ has the same value as $\frac{48}{60}$, which has the same value as $\frac{32}{40}$.
- ✔ $\frac{4}{5} = \frac{4 \times 12}{5 \times 12} = \frac{48}{60}$
- ✔ $\frac{32}{40} = \frac{32 \div 8}{40 \div 8} = \frac{4}{5}$
- ✔ $\frac{32}{40} = \frac{32 \times 1.5}{40 \times 1.5} = \frac{48}{60}$

Not all fractions with large numbers, however, can be changed to smaller numbers. Certain rules have to be followed so that the fraction maintains its integrity; it has to have the same value as it did originally.

To reduce fractions to their lowest terms, follow these steps:

As an example, reduce $\frac{48}{60}$ to lowest terms.

1. **Look for numbers that evenly divide both the numerator and denominator.**

   In the example, 12 goes into both 48 and 60 evenly.

2. **Do the division.**

   $$\frac{48 \div 12}{60 \div 12} = \frac{4}{5}$$

3. **Plug the reduced fraction into your problem.**

   So, instead of working with $\frac{48}{60}$, you can work with $\frac{4}{5}$.

When can you use this reducing process? Well, what if you spent 48 minutes waiting in line to buy your airline ticket? That's 48 minutes out of the total 60 minutes in an hour. As a fraction, that's written $\frac{48}{60}$. You can see that 48 out of 60 is a big hunk of time. To get a better picture of what is going on, put the fraction in lowest terms: 12 divides both 48 and 60 evenly.

You spent $\frac{4}{5}$ of the hour standing in line.

Realizing why smaller is better

Why is $\frac{4}{5}$ better than $\frac{48}{60}$? Most people can relate better to smaller numbers. You can picture four out of five things in your mind more easily than you can picture 48 out of 60 — refer to Figure 3-1 if you don't believe me. A couple more examples may help you get this:

- A survey found that 162 out of 198 people preferred Bix Peanut Butter. The fraction $\frac{162}{198}$ reduces to $\frac{9}{11}$, which offers more information as far as the preference for the peanut butter.

- An ad on TV says, "Nine out of ten dentists surveyed prefer Squishy Toothpaste." I've always wondered how many dentists were actually surveyed. The fraction $\frac{9}{10}$ gives good information as far as the preference, but were only ten dentists surveyed or were a thousand?

Figure 3-1 is an illustration that makes a case for smaller being better. On the left, there are a total of sixty divisions. On the right, there are a total of five divisions.

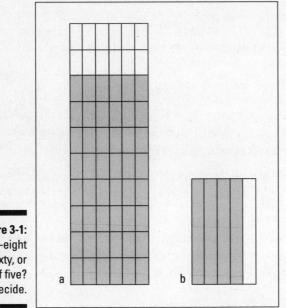

Figure 3-1:
Forty-eight
of sixty, or
four of five?
You decide.

- You paid 18 installments out of a total of 36 for a new television. Both numbers are divisible by 18 — 18 goes into 18 once (1), and 18 goes into 36 twice (2). So you know that you have made one-half $\left(\frac{1}{2}\right)$ of your total payments.

  That's $\frac{18}{36} = \frac{18 \div 18}{36 \div 18} = \frac{1}{2}$.

  You're half done or have half to go — depending on whether you're a glass-half-full or glass-half-empty personality.

✔ Your favorite pitcher has pitched 96 innings so far. Because there are 9 innings in a regulation game, he has pitched $\frac{96 \text{ innings}}{9 \text{ innings}}$ per game = $10\frac{6}{9}$ games.

Because $\frac{96}{9}$ is an improper fraction, first divide 96 by 9 and write the remainder as a fraction. $\frac{96}{9} = 10\frac{6}{9} = 10\frac{2}{3}$ games.

A *remainder* is the value left over when one number is divided by another.

Changing the look without changing the value or what to do when you can't go any lower

It's always nice when you can reduce a fraction to make it more user-friendly. The fraction $\frac{3}{4}$ is much nicer than $\frac{447}{596}$. Sometimes, though, the fraction just doesn't want to cooperate. You still have options: You can round up or down, and then you can multiply or divide by one.

Rounding up or down

Try reducing $\frac{25}{36}$ into its lowest terms. Twenty-five can be divided by 5 or 25, and 36 can be divided by 2, 3, 4, 6, 9, 12, 18 and 36, but none of these match with 5 or 25. Even though 25 and 36 are not prime numbers, they have no common factor. The fraction cannot be reduced.

In cases where the fraction won't reduce to lower terms, you can just leave well enough alone, or, if you're so inclined, round the numerator or denominator up or down to make it reducible.

For example, $\frac{301}{498}$ doesn't reduce. But, if you round the numerator down to 300 and the denominator up to 500, you get an approximate fraction $\frac{300}{500}$ that reduces to $\frac{3}{5}$. In this case, the rounding doesn't change the value by much. You just have to use your judgment.

Dividing by one

Any number divided by one equals that number: For any real number *n*,

$$n \div 1 = n.$$

So, knowing this allows you to change how a fraction looks without changing its value. See how it works in the following example?

$$\frac{8}{12} \div 1 \text{ could be } \frac{8}{12} \div \frac{4}{4} = \frac{2}{3} = \frac{8}{12}$$

You do the same thing on the top and bottom of the fraction, so you really just divided by one, which doesn't change the value — just how it looks.

Multiplying by one

Any number multiplied by one equals that number: For any real number n,

$$n \times 1 = n.$$

Like division, you can multiply by one and change how a fraction looks without changing its value. That is,

$$4 \times 1 = 4 \qquad -8 \cdot 1 = -8 \qquad \frac{3}{4} \times 1 = \frac{3}{4}$$

In the case of fractions, instead of actually using one, a fraction equal to one is used.

$$1 = \frac{3}{3} = \frac{7}{7} = \frac{10}{10} = \cdots$$

Using the fractional value for the number one allows you to change how fractions look without changing their value.

$$\frac{2}{3} \times 1 \text{ could be } \frac{2}{3} \times \frac{4}{4} = \frac{8}{12} = \frac{2}{3}$$

Table 3-1 lists some equivalent fractions of everyday things.

Table 3-1	Some Equivalent Fractions
Fractions	**Equivalent**
$\frac{1}{2} = \frac{2}{4} = \frac{3}{6} = \frac{4}{8} = \frac{5}{10}$	One half of a basketball game
$\frac{2}{3} = \frac{4}{6} = \frac{6}{9} = \frac{8}{12} = \frac{10}{15}$	Two periods of a hockey game
$\frac{4}{7} = \frac{8}{14} = \frac{12}{21} = \frac{16}{28} = \frac{20}{35}$	Four days of a week
$\frac{5}{9} = \frac{10}{18} = \frac{15}{27} = \frac{20}{36} = \frac{25}{45}$	Five innings of a baseball game
$\frac{7}{12} = \frac{14}{24} = \frac{21}{36} = \frac{28}{48} = \frac{35}{60}$	Seven months of a year
$\frac{23}{24} = \frac{46}{48} = \frac{69}{72} = \frac{92}{96} = \frac{115}{120}$	Twenty-three hours of a day

Fitting Fractions Together

To add, subtract, or compare fractions, you need fractions with the same number of equal pieces. In other words, the denominators have to be the same.

Finding common denominators

Common denominators or the *same numbers* in the denominators are necessary for adding, subtracting, and comparing fractions. Carefully selected fractions that equal the number one are used to create common denominators because multiplying by one doesn't change a number's value.

Follow these steps to find a common denominator for two fractions and write the equivalent fractions. Use the two fractions $\frac{7}{18}$ and $\frac{5}{24}$ as an example:

1. **Look to see if you can find a common denominator just by observation.**

   The numbers 18 and 24 are pretty big, and nothing jumps out at first. If you *do* find one, go down to step 4.

2. **Determine which fraction has the larger denominator.**

   In this case the 24 is the larger of the two denominators.

3. **Check to see if the smaller denominator divides the larger one evenly. If not, check multiples of the larger denominator until you find one that the smaller denominator can divide into evenly, too.**

   The number 18 doesn't divide 24 evenly. Two times 24 is 48, but 18 doesn't divide that evenly, either. Three times 24 is 72. Eighteen *does* divide that evenly. The common denominator is 72.

4. **Write the two fractions as equivalent fractions with the common denominator.**

   The number 24 divides 72 three times, so the fraction $\frac{5}{24}$ is multiplied by $\frac{3}{3}$.

   $$\frac{5 \times 3}{24 \times 3} = \frac{15}{72}$$

   Eighteen divides 72 four times, so the fraction $\frac{7}{18}$ is multiplied by $\frac{4}{4}$:

   $$\frac{7 \times 4}{18 \times 4} = \frac{28}{72}$$

Now follow the same steps, using the two fractions $\frac{3}{8}$ and $\frac{5}{12}$ as an example:

1. Find a common denominator.

   The numbers 8 and 12 both have the number 24 as a multiple. If you see this right away, you can go to step 4.

2. Determine which fraction has the larger denominator.

   In this case, the 12 is the larger of the two denominators.

3. Check multiples.

   Eight doesn't divide 12 evenly. Two times 12 is 24, and 8 does divide that evenly. The common denominator is 24.

4. Write as equivalent fractions.

   Twelve divides 24 two times, so the fraction $\frac{5}{12}$ is multiplied by $\frac{2}{2}$.

   $$\frac{5 \times 2}{12 \times 2} = \frac{10}{24}$$

   Eight divides 24 three times, so the fraction $\frac{3}{8}$ is multiplied by $\frac{3}{3}$.

   $$\frac{3 \times 3}{8 \times 3} = \frac{9}{24}$$

Sometimes you can get a denominator that fits quickly by multiplying the two denominators together. This method doesn't always give the best or smallest choice, but it's efficient.

Find a common denominator for $\frac{5}{8}$ and $\frac{4}{9}$. Multiply the two denominators together, $8 \times 9 = 72$. Then you can see that you want:

- $\frac{5}{8} = \frac{?}{72}$ and $\frac{4}{9} = \frac{?}{72}$.

  By multiplying the numerators by the same factors, you get:

  $\frac{5 \times 9}{8 \times 9} = \frac{45}{72}$ and $\frac{4 \times 8}{9 \times 8} = \frac{32}{72}$.

- Find a common denominator for $\frac{3}{5}$ and $\frac{5}{12}$.

  Multiply 5×12 to get a common denominator of 60.

  $\frac{3}{5} = \frac{36}{60}$ and $\frac{5}{12} = \frac{25}{60}$.

Working with improper fractions

Multiplying and dividing improper fractions (see the "Getting to know improper fractions" section earlier in this chapter for an introduction) is no more difficult than multiplying or dividing other fractions. Understanding the final result is easier, however, if you write the answer as a mixed number (see the "Mixing it up with mixed numbers" section earlier in this chapter).

To change an improper fraction to a mixed number, divide the numerator by the denominator. The number of times the denominator divides is the whole number in front, and the remainder — the leftover value — is written as a proper fraction, which has a numerator smaller than the denominator. See the similarity and difference between mixed numbers and improper fractions in the following examples:

- $\frac{11}{9} = 1\frac{2}{9}$: The number 9 divides 11 once with 2 left over.

- $\frac{26}{7} = 3\frac{5}{7}$: The number 7 divides 26 three times with 5 left over.

- $\frac{402}{11} = 36\frac{6}{11}$: Eleven divides 402 thirty-six times with 6 left over. This example makes it especially apparent that the mixed number is more understandable.

Putting Fractions to Task

Now that you know everything about fractions — their proper names, charac-ter-istics, strong and weak points, and so on — it's time to put them to work. The rules for addition, subtraction, multiplication, and division of fractions are the same ones used when variables are added. This is reassuring! The rules don't change.

Adding and subtracting fractions

Adding and subtracting fractions takes a little special care. You can add quarts and gallons if you change them to the same unit (quarts). It's the same with fractions. You can add thirds and sixths if you find the common denominator first.

To add or subtract fractions:

1. **Convert the fractions so that they have the same number in the denominators.**

   Find out how to do this in the "Finding common denominators" section.

2. **Add or subtract the numerators. Leave the denominators alone.**

3. **Reduce the answer, if needed.**

A question about an athlete named Jim can demonstrate:

✔ Jim played for half an hour in yesterday's soccer game and for 20 min-utes in today's game. How long did Jim play altogether?

Set up a simple equation, such as $\frac{1}{2} + \frac{1}{3}$ = hours Jim played.

One-half and one-third don't fit together. You can't just add the numera-tors and the denominators because two-fifths doesn't make any sense $\left(\frac{1}{2} + \frac{1}{3} \text{ does not equal } \frac{2}{5}\right)$. But, you realize that $\frac{1}{2}$ can also be $\frac{2}{4}, \frac{3}{6}, \frac{4}{8}, \frac{5}{10}$, or many other things, and $\frac{1}{3}$ can also be $\frac{2}{6}, \frac{3}{9}, \frac{4}{12}, \frac{5}{15}$, or many other things.

So, you can fit the two fractions together by multiplying both the numer-ators and the denominators by the same number: Multiply $\frac{1}{2}$ by $\frac{3}{3}$ to make $\frac{3}{6}$; multiply $\frac{1}{3}$ by $\frac{2}{2}$ to make $\frac{2}{6}$. $\left(\frac{3}{3} = 1 = \frac{2}{2}\right)$

Notice that $\frac{1}{2}$ and $\frac{1}{3}$ each can have a 6 for the denominator, so that they can fit together: $\frac{1}{2}$ and $\frac{1}{3}$ are equal to $\frac{3}{6}$ and $\frac{2}{6}$.

Now you can add the numerators $\frac{3}{6}$ and $\frac{2}{6}$.

Then solve the equation: $\frac{1}{2} + \frac{1}{3} = \frac{3}{6} + \frac{2}{6} = \frac{3+2}{6} = \frac{5}{6}$.

Jim played $\frac{5}{6}$ of an hour altogether.

Another real-life situation shows you how you can make fractions fit to do a simple subtraction.

✔ In her will, Jane gave $\frac{4}{7}$ of her money to the Humane Society and $\frac{1}{3}$ of her money to other charities. How much was left for her children's inheritance?

The fractions $\frac{4}{7}$ and $\frac{1}{3}$ aren't compatible. You can't combine or compare them. The fraction $\frac{4}{7}$ can be $\frac{8}{14}$ or $\frac{12}{21}$ or $\frac{16}{28}$ and more. The fraction $\frac{1}{3}$ can be $\frac{2}{6}, \frac{3}{9}, \frac{4}{12}, \frac{5}{15}, \frac{6}{18}, \frac{7}{21}$ and more.

It may take a while to find a good fit, but $\frac{4}{7} = \frac{12}{21}$ and $\frac{1}{3} = \frac{7}{21}$.

Add the numerators to get the total designation to charity in Jane's will.

$$\frac{12}{21} + \frac{7}{21} = \frac{19}{21}$$

Subtract that total from the whole of Jane's proceeds to find what portion is allotted to her children.

$$\frac{21}{21} = \frac{19}{21} = \frac{2}{21}$$

Jane's children will be awarded $\frac{2}{21}$ of Jane's estate.

Multiplying fractions

Multiplying fractions is a tad easier than adding or subtracting them. This is because you don't need to find a common denominator first. The only catch is that you have to change any mixed numbers to improper fractions. Then, at the end, you may have to change the fraction back again to a mixed number.

When multiplying fractions follow these steps:

1. **Change all mixed numbers to improper fractions.**

2. **Multiply the numerators together and the denominators together.**

3. **Reduce the answer if necessary.**

Fred and Sadie's stories offer opportunities to multiply fractions:

✔ Fred ate $\frac{2}{3}$ of a $\frac{3}{4}$-pound box of candy. How much candy did he eat?

$\frac{2}{3} \times \frac{3}{4} = \frac{6}{12} = \frac{1}{2}$ pound of candy (and 6 zillion calories).

✔ Sadie worked $10\frac{2}{3}$ hours at time-and-a-half. How many hours will she get paid for?

$10\frac{2}{3} \times 1\frac{1}{2} = \frac{32}{3} \times \frac{3}{2} = \frac{96}{6} = 16$ earned hours to multiply by the hourly rate.

Reducing the fractions *before* multiplying can make multiplying fractions easier. Smaller numbers are more manageable, and if you reduce the fractions before you multiply, you don't have to reduce them afterwards.

This is another way of looking at Fred's candy problem:

✔ The expression $\frac{2}{3} \times \frac{3}{4}$ has a 2 in the first numerator and a 4 in the second denominator. Even though they aren't in the same fraction, this is a multiplication problem. Multiplication is *commutative,* meaning that it doesn't matter what order you multiply the numbers: So you can pretend that the 2 and 4 are in the same fraction.

So, dividing the first numerator by 2 and the second denominator by 2, you get

$$\frac{2}{3} \times \frac{3}{4} = \frac{1}{3} \times \frac{3}{2}$$

But $\frac{1}{3} \times \frac{3}{2}$ has a 3 in the first denominator and a 3 in the second numerator. You can divide by 3!

So $\frac{1}{3} \times \frac{3}{2} = \frac{1}{1} \times \frac{1}{2} = \frac{1}{2}$, which is the same answer as in the original example.

In the previous example either method — reducing before or after multiplying — was relatively easy. This example shows how necessary reducing *before* working the problem can be.

✔ Multiply the two fractions: $\frac{360}{121} \times \frac{77}{900}$

The numerator of the first fraction and the denominator of the second fraction can each be divided by 180.

$$\frac{360}{121} \times \frac{77}{900} = \frac{2}{121} \times \frac{77}{5}$$

The denominator of the first fraction and the numerator of the second fraction each can be divided by 11.

$$\frac{2}{121} \times \frac{77}{5} = \frac{2}{11} \times \frac{7}{5}$$

Now the multiplication is simple:

$$\frac{2}{11} \times \frac{7}{5} = \frac{14}{55}$$

This is much simpler than the original problem would have been!

The operations of addition and multiplication have another special feature that subtraction and division don't have. You can perform the operation on more than two fractions at a time.

✔ The following example shows how to multiply three fractions together. A situation such as this could happen if you were applying one discount after another to an original list price.

$$\frac{5}{6} \times \frac{3}{8} \times \frac{4}{7} = \frac{5 \times 3 \times 4}{6 \times 8 \times 7} = \frac{60}{336} = \frac{5}{28}$$

You can make this easier if you reduce first: the 4 and 8 in the third numerator and second denominator, and the 3 and 6 in the second numerator and first denominator.

✔ This example involves mixed numbers:

$$3\frac{1}{3} \times 5\frac{1}{4} \times 2 = \frac{10}{3} \times \frac{21}{4} \times \frac{2}{1}$$

Reducing first would mean dividing by 3 and dividing by 2.

$$\frac{10}{3} \times \frac{21}{4} \times \frac{2}{1} = \frac{10}{1} \times \frac{7}{2} \times \frac{1}{1} = \frac{70}{2} = 35$$

Dividing fractions

Dividing fractions is as easy as pie! That is, dividing the leftover pie into enough pieces so that everybody at your table gets a piece. Actually, it's just like multiplying fractions, except that the numerator and the denominator of the second fraction change places.

When dividing fractions:

1. **Change all mixed numbers to improper fractions.**

2. **Flip the second fraction, placing the bottom number on top and the top number on the bottom.**

3. **Continue as with the multiplication of fractions.**

The following example shows how this system works:

✔ You bought $6\frac{1}{2}$ pounds of sirloin steak and want to cut it into pieces that weigh $\frac{3}{4}$ of a pound each.

$$6\frac{1}{2} \div \frac{3}{4} = \frac{13}{2} \div \frac{3}{4} = \frac{13}{2} \times \frac{4}{3} = \frac{52}{6} = 8\frac{4}{6} = 8\frac{2}{3}$$

Having $8\frac{2}{3}$ pieces means that you can cut the steak into 8 pieces that each weigh a full $\frac{3}{4}$ pound with a small piece left over.

Dealing with Decimals

Decimals are nothing more than glorified fractions. They're special because their denominators are always 10, 100, 1,000, and so on — powers of ten. Because they're so special, you don't even have to bother with the denominator part. Just write the numerator and use a decimal point to indicate that it's really a fraction.

Decimal point abuse

When a decimal point is misused, it can be costly. Ninety-nine cents can use a cent symbol, 99¢, or a dollar symbol, $.99. It's when people aren't careful or don't understand that you see .99¢. You figure that they *mean* 99 cents, but that's not what this says. The price .99¢ means ninety-nine hundredths of a cent — not quite a cent.

A friend of mine once challenged a hamburger establishment on this. They advertised a super-duper hamburger for the regular price and any

additional for .99¢. He went in and asked for his regular-priced hamburger and two additional for one cent each. (He was willing to round up to a whole penny.) When the flustered clerk finally realized what had happened, he honored my friend's request. Actually, the friend wouldn't have made a big deal of it. He just wanted to make a point. But you can bet that the sign was corrected quickly.

The decimal point symbol looks the same as a period at the end of a sentence, which is a really small figure, so you'll have to watch carefully for it. More importantly, though, you need to watch where the decimal point is *placed* in the number.

The number of decimal places to the right of the decimal point tells you the number of zeros in the power of ten that is written in the denominator.

Check out the decimal placement in the following examples:

- .3 has just one digit (3) to the right of the decimal point. The .3 is $\frac{3}{10}$.
- .408 has three digits (408) to the right of the decimal point. The .408 is $\frac{408}{1,000}$.
- 60.0003 has four digits (0003) to the right of the decimal point. The 60.0003 is 60 and $\frac{3}{10,000}$.

A *digit* is any single number from zero through nine.

Decimal fractions are great because you can add, subtract, multiply, and divide them so easily. That's why it's often desirable to change a fraction to a decimal.

Changing fractions to decimals

All fractions can be changed to decimals. In Chapter 1, I tell you that rational numbers have decimals that can be written exactly as fractions. The decimal forms of rational numbers either end or repeat a pattern. Here is how you can change the fractions to decimals.

To change a fraction to a decimal, just divide the top by the bottom.

$\frac{3}{4}$ becomes $4\overline{)3.00}$ = .75 so $\frac{3}{4}$ = .75

$\frac{15}{8}$ becomes $8\overline{)15.000}$ = 1.875 so $\frac{15}{8}$ = 1.875

If the division doesn't come out evenly, you can stop after a certain number of decimal places and *round off*.

Rounding numbers:

1. **Determine the number of *places* you want and go one further.**

2. **Increase the last place you want by one number if the *one further* is five or bigger.**

3. **Leave the last place you want as is, if the *one further* is less than five.**

The fraction $\frac{5}{9}$ won't divide evenly, and it'll go on forever and ever when divided. So divide and decide when to stop.

Change $\frac{5}{9}$ to a decimal.

$9\overline{)5.000000\ldots}$ = .5555\ldots

If you choose to round to 3 decimal places, $\frac{5}{9}$ = .55555 . . . ≈.556

The symbol ≈ means *approximately the same* or *about equal*. This is a useful symbol to use when rounding a number.

✔ Now try changing $1\frac{5}{8}$ to a decimal.

$1\frac{5}{8}$ = $\frac{13}{8}$

$8\overline{)13.000}$ with quotient 1.625

✔ Try changing $\frac{4}{7}$ to a decimal.

$7\overline{)4.000000000000\ldots}$ with quotient .571428571428\ldots

If this is rounded to 4 places, the answer is $\frac{4}{7}$ ≈ .5714

If this is rounded to 5 places, the answer is $\frac{4}{7}$ ≈ .57143

Changing decimals to fractions

To change a decimal into a fraction, put the numbers to the right of the decimal point in a numerator. Put the number one in the denominator followed by as many zeros as the numerator has digits. Reduce the fraction if necessary.

✔ Change .36 into a fraction:

$$.36 = \frac{36}{100} = \frac{9}{25}$$

There were two digits in 36, so the 1 is followed by two zeros.

✔ Change .403 into a fraction:

$$.403 = \frac{403}{1,000}$$

There were three digits in 403, so the 1 is followed by three zeros.

✔ Change .0005 into a fraction:

$$.0005 = \frac{5}{10,000} = \frac{1}{2,000}$$

Don't forget to count the zeros in front of the 5 when counting the number of digits.

✔ Change 3.025 into a fraction:

$$3.025 = 3\frac{25}{1,000} = 3\frac{1}{40}$$

You just need to be able to count decimal places and zeros.

Chapter 4

Exploring Exponents and Raising Radicals

*E*xponents, those small symbols, slightly higher and to the right of numbers, were developed so that mathematicians wouldn't have to keep repeating themselves! What is an exponent? An *exponent* is the small, super-scripted number to the upper right of the larger number that tells you how many times you multiply the larger number, called the *base,* by itself. That is, three to the fourth power (3^4) is three multiplied by itself four times. Got that? Now, try it and see what happens.

$$3^4 = 3 \times 3 \times 3 \times 3 = [(3 \times 3) \times 3] \times 3 = 81$$

So, really, three to the fourth power (3^4) is another way of saying 81.

When you run into grouping symbols, like parentheses or brackets, do what is in the grouping symbol first, before doing any other operations. For more information on grouping symbols, refer to Chapter 1.

Paying off a royal debt exponentially

There's an old story about a king who backed out on his promise to the knight who saved his castle from a fire-breathing dragon. The king was supposed to pay the knight two bags of gold for his bravery and for the successful endeavor.

After the knight had slain the dragon, the king was reluctant to pay up — after all, no more fire breathing in the neighborhood! So the frustrated knight, wanting to get his just reward, struck a bargain with the king: On January 1st, the king would pay him one pence, and he would double the amount every day until the end of April. So, on January 2nd, the king would pay him two pence. On January 3rd, the king would pay him four pence. On January 4th, the king would pay him eight pence. On January 5th, the king would pay him 16 pence. And this would continue through April 30th. The king thought that this was a pretty good deal. After all, the knight was just asking for some of the smallest coins that the king had. So he agreed and started paying off the knight. It went pretty well until the end of January. On January 20th, he had to pay 524,288 pence. Then, on February 20th he had to pay 109,951,162,800 pence. On the last day, April 30th, he had to pay over 66,461,399,790,000,000,000,000,000,000,000 pence. Add up all the pence on all the days and the total amount was more than 132,922,799,600,000,000,000,000,000,000,000 pence.

If a *pence* is close to a penny, then this is over a trillion trillion dollars.

Guess who was king *then!*

Multiplying the Same Thing Over and Over and . . .

When algebra was first written with symbols — instead of with all words — there were no exponents. If you wanted to multiply the variable *y* times itself six times, you'd write it: *yyyyyy*. (Kinda like talking to a three-year-old, "Why, why, why, why, why, why?") Writing the variable over and over can get tiresome (just like three-year-olds), so the wonderful system of exponents was developed.

A *variable* is a letter that represents an unknown number or what you're solving for in an algebra problem.

Powering up exponential notation

It's one thing to write numbers with exponents and another to know what these exponents mean and what you can do with them. Using exponents is so convenient that it's worth the time and trouble to find out the rules for using them.

An *exponent* is a small number written above and to the right of the *base* — the number you're multiplying times itself. The exponent is usually in smaller print than the base. The base can be any real number (check the glossary if you need a definition). The exponent, the power, can be any real number, also. An exponent can be positive or negative or fractional or even a radical. What power! The following example gives you another demonstration of how convenient exponents can be.

$$x^n = x \cdot x \cdot x \cdot x \cdot x \cdots n \text{ times}$$

x: the base   *x* can be any real number.

n: the power, or exponent   *n* can be any real number.

Even though the *x* in the expression x^n can be any real number and the *n* can be any real number, they can't both be 0 at the same time. 0^0 really has no meaning in algebra. It takes a calculus course to discuss this. Also, if *x* is equal to 0, then *n* can't be negative.

In the examples below, the base is multiplied *n* times, and the exponential expression is evaluated.

$$2^4 = 2 \cdot 2 \cdot 2 \cdot 2 = 16$$
$$3^5 = 3 \cdot 3 \cdot 3 \cdot 3 \cdot 3 = 243$$
$$10^8 = 10 \cdot 10 \cdot 10 \cdot 10 \cdot 10 \cdot 10 \cdot 10 \cdot 10 = 100,000,000$$

The nice thing about exponents of ten is that the power tells you how many zeros are in the answer.

In this example, several bases are multiplied together. Each base has its own, separate exponent. The *x, y,* and *z* represent real numbers.

$$3^3 x^2 y^4 z^6 = 3 \cdot 3 \cdot 3 \cdot x \cdot x \cdot y \cdot y \cdot y \cdot y \cdot z \cdot z \cdot z \cdot z \cdot z \cdot z$$

You can see why it's preferable to use the powers. And in the next example, the base is actually a binomial. The parenthesis means that you add the two values together before applying the exponent.

$$(a + b)^3 = (a + b) \cdot (a + b) \cdot (a + b)$$

Comparing with exponents

It's easier to compare amounts when you use exponents. Try to compare two numbers: 943,260,000,000,000,000,000,000 and 8,720,000,000,000,000,000,000,000.

The first number may look bigger because of the first three digits, but this is deceiving. To discover the real value of a large number follow these steps. Write these using multiplication and exponents:

1. **Write the number(s) as a number between 1 and 10 times a power of ten.**

   Using the previous numbers, you get

   $$943{,}260{,}000{,}000{,}000{,}000{,}000{,}000 =$$
   $$9.4326 \times 100{,}000{,}000{,}000{,}000{,}000{,}000{,}000$$

   and

   $$8{,}720{,}000{,}000{,}000{,}000{,}000{,}000{,}000 =$$
   $$8.72 \times 1{,}000{,}000{,}000{,}000{,}000{,}000{,}000{,}000$$

2. **Write each power of ten as an exponential expression with the power indicating the number of zeros.**

   In the example, this translates to

   $$943{,}260{,}000{,}000{,}000{,}000{,}000{,}000 = 9.4326 \times 10^{23}$$

   $$8{,}720{,}000{,}000{,}000{,}000{,}000{,}000{,}000 = 8.72 \times 10^{24}$$

3. **Compare the numbers.**

   The number with the higher power of ten is the larger number. If the powers are the same, then compare the numbers multiplying the power of ten.

   $$8.72 \times 10^{24} > 9.4326 \times 10^{23}$$

Why is the number with the higher power of ten larger? Look at these two numbers that are a little more manageable (they don't have over twenty zeros).

Compare 8×10^{2} and 9×10^{1}. That's comparing $8 \times 100 = 800$ with $9 \times 10 = 90$. Even though the 9 is bigger than the 8, it's the larger power of ten that "wins."

FUN FACT

Exponents

The first exponents appeared in about 1636 when the base was written on a regular line and the exponent was elevated a little to the right. The first exponents were expressed in Roman numerals, so y cubed would be written y^{III}. There were many who resisted using these, at first, and continued to write y cubed as *yyy*. Rene Descartes tended not to use two as an exponent. He still preferred to write *aa* rather than a^{2}. Isaac Newton is credited with being the first to recognize and use negative and fractional exponents.

Taking notes on scientific notation

Large numbers are used when talking about distances between planets, the number of grains of sand, or the amount of money spent by the government. Very small numbers are used for measurements of plant or animal cells, the size of atoms, or other such teeny things. *Scientific notation* is a standard way of recording these very large and very small numbers so they can fit on one line in the page of a book and so they can be compared more easily. Computations with them are easier in this form, too.

The form for scientific notation is: $N \times 10^a$ where N is a number between 1 and 10 (but not 10 itself), and where *a* is an integer (positive or negative number).

You can write large and small numbers in scientific notation by moving the decimal point until the new form is a number from one up to ten, and then indicating how many places the decimal point was moved by the power you raise 10 to.

Whether the power of 10 is positive or negative depends on whether you move the decimal to the right or to the left. Moving the decimal to the right makes the exponent negative, moving it to the left gives you a positive exponent. The following examples show you how this works.

- $41,000 = 4.1 \times 10^4$

  Move the decimal place four spaces to the left, and, conveniently, you raise 10 to the fourth power.

- $312,000,000,000 = 3.12 \times 10^{11}$

  The decimal place is moved 11 spaces to the left.

- $.00000031 = 3.1 \times 10^{-7}$

  The decimal place is moved seven spaces to the *right* this time. This is a very *small* number, and the exponent is negative.

- $.2 = 2 \times 10^{-1}$

  The decimal place is moved one space to the right.

Exploring Exponential Expressions

Expressing very large numbers or very small numbers exponentially makes them so much easier to deal with! This is also true when studying situations that involve doing the same thing over and over again.

Picture a cat stalking a mouse. They're about 100 inches apart. Every time the mouse starts nibbling at the hunk of cheese, the cat takes advantage of the mouse's distraction and creeps closer by one-tenth the distance between them. The cat wants to get about six inches away — close enough to pounce. How far apart are they after four moves? How about after ten moves? How long will it take before the cat can pounce on the mouse?

Use these steps to stalk your own mouse (or to figure any decreasing distance):

1. **Express the incremental move as a fraction.**

   In the sample problem, that's easy because the cat creeps closer by one-tenth the distance between them.

2. **Multiply the total distance by the fraction to get the length of the move.**

   The cat and mouse are 100 inches apart, so you multiply 100 times $\frac{1}{10}$ to get 10 inches.

3. **Subtract the length of the move from the current distance.**

   100 inches minus 10 inches leaves 90 inches between them.

4. **Multiply the current distance by the fraction to find the distance of the second move.**

   Second move: Multiply 90 times $\frac{1}{10}$ to get 9 inches.

5. **Subtract the length of the move from the current distance.**

   90 inches minus 9 inches leaves 81 inches between them.

6. **Multiply the current distance by the fraction to find the distance of the third move.**

   Third move: Multiply 81 times $\frac{1}{10}$ to get 8.1 inches.

7. **Subtract the length of the move from the current distance.**

   81 inches minus 8.1 inches = 72.9 inches between them.

And so on, and so on, and so on. (Aren't you glad the cat wasn't 200 inches away?)

However, there is an easier way. Instead of finding one-tenth the distance remaining each time and subtracting, switch to finding the distance remaining between them, which is nine-tenths of the distance before that move. One-tenth plus nine-tenths equals one — the whole amount.

In each step, you multiply by $\frac{9}{10}$ — the fraction of the distance left after the move times the current distance. Nine-tenths times the current distance = new distance. Then there's just one operation to deal with each time.

1. Find the distance left between them after the first move by multiplying the current distance by $\frac{9}{10}$.

   $\frac{9}{10}$ of 100 = $\frac{9}{10} \times 100$ = 90 inches between them.

2. Find the distance left between them after the second move by multiplying the current distance by $\frac{9}{10}$.

   $\frac{9}{10}$ of 90 = $\frac{9}{10} \times 90$ = 81 inches between them.

3. Find the distance left between them after the third move by multiplying the current distance by $\frac{9}{10}$.

   $\frac{9}{10}$ of 81 = $\frac{9}{10} \times 81$ = 72.9 inches between them.

4. Find the distance left between them after the fourth move by multiplying the current distance by $\frac{9}{10}$.

   $\frac{9}{10}$ of 72.9 = $\frac{9}{10} \times 72.9$ = 65.61 inches between them.

Again, as you see, this can get pretty tedious. The best way to find the answer is to use exponents. Figuring this problem using powers, or exponents, can make the computation easier. The third time is the charm for finding the distance between the cat and the mouse. Just follow these steps:

Distance to pounce = $100\left(\frac{9}{10}\right)^n$ where *n* is the number of moves the cat has made.

Perform the operations inside the grouping symbol first.

In this formula, because the fraction $\frac{9}{10}$ is inside parenthesis, apply the exponent just outside the parenthesis to the fraction first. Multiply the fraction *n* times itself before multiplying it by 100.

  ✔ Find the distance between the cat and the mouse using this formula:

   After the third move, the distance between them is $100\left(\frac{9}{10}\right)^3$ = 72.9 inches

   After the tenth move, the distance between them is $100\left(\frac{9}{10}\right)^{10} \approx 34.87$ inches. This still isn't close enough to pounce. I'm using the approximately symbol, $\approx$, here because the actual answer has many more decimal places and you don't need all that information.

   After the twenty-sixth move, the distance between them is $100\left(\frac{9}{10}\right)^{26} \approx 6.46$ inches

It'll take one more move to be within the six-inch *pounce* distance. Do you suppose the mouse still hadn't caught on after 26 moves? If not, then it deserves to be pounced upon.

Some other quick examples:

  ✔ To find the population of a city that is growing by 5 percent per year if it had 10,000 people in 1990. The equation needed is:

   Population = $10,000(1.05)^n$. Let *n* be the number of years since 1990.

In 1995, $n = 5$, so the population is $= 10,000(1.05)^5 \approx 12,763$.

Round this to a whole number and use the approximately-equal-to sign so that there won't be a "piece" of a person.

✔ You want to find out the total distance (up and down and up and down and up . . .) that a super ball travels in n bounces if it always bounces back 75 percent of the distance it falls. You're going to drop the super ball from a window that is 40 feet high onto a nice, smooth sidewalk.

The equation you need is:

$$Distance = 40 + 240\left[1 - .75^n\right].$$

Let n be the number of bounces. Then the total distance can be found with the formula.

After one bounce and before the second one, the total distance is 40 feet + 30 feet + 30 feet = 100 feet.

Check this with the formula:

$$Distance = 40 + 240\left[1 - .75^1\right] = 100\, feet.$$

After ten bounces, the total distance is 40 feet + 30 feet + 30 feet + 22.5 feet + . . . UGH! Use the formula!

$$Distance = 40 + 240\left[1 - .75^{10}\right] \approx 266.48\ feet.$$

Here you have just the formula without all the basic steps needed to work it out. But you can see how it's possible to solve such problems. It takes just a bit more information on sequences and series to set it up, but you can do this algebra part. Go to the *Bouncing Ball* link at `http://hilltop.bradley.edu/%7Esterling/facinfo.html` for more help.

Multiplying Exponents

You can multiply many exponential expressions together without having to change their form into the big or small numbers they represent. The only requirement is that the bases of the exponential expressions have to be the same. The answer is then a nice, neat exponential expression.

x^n: x is the base; it can be any real number. n is the power or exponent; it, too, can be any real number. To multiply two of these types of numbers together, the bases must be the same value. So, you *can* multiply $2^4 \cdot 2^6$ and $a^6 \cdot a^8$, but you *cannot* multiply $3^5 \cdot 4^5$ because the bases are not the same (although the exponents are).

To multiply powers of the same base, add the exponents together:

$$x^a \cdot x^b = x^{a+b}.$$

Check out these examples:

- $2^4 \cdot 2^9 = 2^{4 + 9} = 2^{13}$
- $a^5 \cdot a^8 = a^{13}$
- $4^a \cdot 4^2 = 4^{a + 2}$

If there's more than one base in an expression with powers, you can combine the numbers with the same bases, find the values, and then write them all together, as the following examples show:

- $3^2 \cdot 2^2 \cdot 3^3 \cdot 2^4 = 3^{2 + 3} \cdot 2^{2 + 4} = 3^5 \cdot 2^6$
- $4x^6 y^5 x^4 y = 4x^{6 + 4} y^{5 + 1} = 4x^{10} y^6$

When there's no exponent showing, such as with y, you assume that the exponent is 1, so in this example write y^1.

Dividing and Conquering

You can divide exponential expressions, leaving the answers as exponential expressions, as long as the bases are the same. Division is the opposite of multiplication, so it makes sense that because you add exponents when multiplying numbers with the same base, you *subtract* the exponents when dividing numbers with the same base. Easy enough?

To divide powers with the same base, subtract the exponents: $x^a \div x^b = x^{a - b}$ where x can be any real number except zero; you can't divide by zero.

These examples show how the division rule works:

- $2^{10} \div 2^4 = 2^{10 - 4} = 2^6$

   These exponentials represent the problem $1{,}024 \div 16 = 64$. It's much easier to leave these as exponents.

- $\dfrac{4x^6 y^3 z^2}{2x^4 y^3 z} = 2x^{6 - 4} y^{3 - 3} z^{2 - 1} = 2x^2 y^0 z^1 = 2x^2 z$:

   The variables represent numbers, so writing this out the long way would be

   $$\frac{2 \cdot 2 \cdot x \cdot x \cdot x \cdot x \cdot x \cdot x \cdot y \cdot y \cdot y \cdot z \cdot z}{2 \cdot x \cdot x \cdot x \cdot x \cdot y \cdot y \cdot y \cdot z}.$$

   By crossing out the common factors, all that's left is $2x^2 z$. Need I say more?

What's this with the exponent of 0 on the y? Read on.

Testing the Power of Zero

If x^3 means $x \cdot x \cdot x$, what does x^0 mean? Well, it doesn't mean x times zero, so the answer isn't zero. x represents some unknown real number; it just can't be zero. To understand how this works, use the following rule for division of exponential expressions involving zero.

Any number to the power of zero equals one, as long as the base number is not zero.

For example, to divide $2^4 \div 2^4$, use the rule for dividing exponential expressions, which says that if the base is the same, subtract the two exponents in the order that they're given. Doing this you find that the answer is $2^{4-4} = 2^0$. But $2^4 = 16$, so $2^4 \div 2^4 = 16 \div 16 = 1$. That means that $2^0 = 1$. This is true of all numbers that can be written as a division problem, which means that it's true for all numbers except those with a base of zero.

See how this power of zero works:

- $m^2 \div m^2 = m^{2-2} = m^0 = 1$
- $4x^3 y^4 z^7 \div 2x^3 y^3 z^7 = 2x^{3-3} y^{4-3} z^{7-7} = 2x^0 y^1 z^0 = 2y$
- $\dfrac{(2x^2 + 3x)^4}{(2x^2 + 3x)^4} = (2x^2 + 3x)^{4-4} = (2x^2 + 3x)^0 = 1$

Notice that the x and z, with their zero exponents, then become ones. And when you multiply by one, the value is unchanged.

Working with Negative Exponents

Negative exponents are a neat little creation. They mean something very specific and have to be handled with care, but they are oh, so convenient to have.

You can use a negative exponent to write a fraction without writing a fraction! It's a way to combine expressions with the same base, whether the different factors are in the numerator or denominator or whatever. It's a way to change division problems into multiplication problems.

Calculating interest

There was big-time lending and borrowing going on in ancient times. In Mesopotamia, bankers used exponential tables to determine how long it would take for money to double at an interest rate of 20 percent annually. Wouldn't it be nice to find a nice Mesopotamian bank to put your savings into? On the other hand, I wouldn't want to take out a loan!

Negative exponents are a way of writing powers of fractions or decimals without using the fraction or decimal. For instance, instead of writing $\left(\frac{1}{10}\right)^{14}$, you can write 10^{-14}.

REMEMBER

A *reciprocal* of a number is the multiplicative inverse of the number. The product of a number and its reciprocal is equal to one.

ALGEBRA RULES
$\frac{1}{+1}$
$\frac{}{2}$

The reciprocal of x^a is $\frac{1}{x^a}$ which can be written as x^{-a}. The variable x is any real number except zero, and a is any real number. And, going to the negative side, $x^{-a} = \frac{1}{x^a}$.

The following examples show you how to change from positive to negative exponents, and vice versa.

> ✔ $2^{-3} = \frac{1}{2^3} = \frac{1}{8}$
>
> The reciprocal of 2^3 is $\frac{1}{2^3} = 2^{-3}$.
>
> ✔ $z^{-4} = \frac{1}{z^4}$
>
> The reciprocal of z^4 is $\frac{1}{z^4} = z^{-4}$. In this case, z cannot be 0.
>
> ✔ $6^{-1} = \frac{1}{6}$
>
> The reciprocal of 6 is $\frac{1}{6} = 6^{-1}$.

But what if you start out with a negative exponent in the denominator? What happens then? The reciprocal of $\frac{1}{3^{-4}}$ is 3^4. Here you start with a negative exponent because $\frac{1}{3^4} = 3^{-4}$. The reciprocal is $\frac{1}{3^{-4}} = 3^4$.

So the negative exponent in the denominator comes up to the numerator with a change in the sign to a positive exponent. Here are two more examples:

> ✔ $\frac{x^2 y^3}{3z^{-4}} = \frac{x^2 y^3 z^4}{3}$
>
> ✔ $\frac{4a^3 b^5 c^6 d}{a^{-1} b^{-2}} = 4a^3 a^1 b^5 b^2 c^6 d = 4a^4 b^7 c^6 d$

Powers of Powers

Because exponents are symbols for repeated multiplication, one way to write $\left(x^3\right)^6$ is $x^3 \cdot x^3 \cdot x^3 \cdot x^3 \cdot x^3 \cdot x^3$. Using the multiplication rule where you just add all the exponents together, you get $x^{3+3+3+3+3+3} = x^{18}$. Wouldn't it be just grand if the rule for raising a power to a power was just to multiply the two exponents together? Lucky you!

ALGEBRA RULES
$\frac{1}{+1}$
$\frac{}{2}$

Raising a power to a power: $\left(x^n\right)^m = x^{n \cdot m}$. When the whole expression, x^n, is raised to the mth power, the new power of x is determined by multiplying n and m together.

These examples show you how raising a power to a power works.

- $\left(6^{-3}\right)^4 = 6^{-3 \cdot 4} = 6^{-12} = \dfrac{1}{6^{12}}$

- $(3^2)^{-5} = 3^{2(-5)} = 3^{-10} = \dfrac{1}{3^{10}}$

- $(x^{-2})^{-3} = x^{(-2)(-3)} = x^6$

- $(3x^2 y^3)^2 = 3^2 x^{2 \cdot 2} y^{3 \cdot 2} = 9x^4 y^6$

  (Each factor in the parenthesis is raised to the power outside the parenthesis.)

- $(3x^{-2} y)^2 (2xy^{-3})^4 = (3^2 x^{-2 \cdot 2} y^{1 \cdot 2})(2^4 x^{1 \cdot 4} y^{-3 \cdot 4}) = \left(9x^{-4} y^2\right)\left(16x^4 y^{-12}\right)$

  $= 144x^0 y^{-10} = \dfrac{144}{y^{10}}$

Notice that the order of operations is observed here. First you raise the expressions in the parentheses to their powers. Then multiply the two expressions together. You get to see multiplying exponents (raising a power to a power) and adding exponents (multiplying same bases). Next is an example with negative exponents.

$$(x^2 y^3)^{-2} (x^{-2} y^{-3})^{-4} = (x^{2(-2)} y^{3(-2)})(x^{(-2)(-4)} y^{(-3)(-4)})$$

$$= (x^{-4} y^{-6})(x^8 y^{12}) = x^{-4+8} y^{-6+12} = x^4 y^6$$

Squaring Up to Square Roots

When you do square roots, the symbol for that operation is a *radical*, $\sqrt{}$.

The radical is a nonbinary operation (involving just one number) that asks you, "What number times itself gives you this number under the radical?" Another way of saying this is if $\sqrt{a} = b$, then $b^2 = a$.

Finding square roots is a relatively common operation in algebra, but working with and combining the roots isn't always so clear.

Expressions with radicals can be multiplied or divided as long as the root power *or* value under the radical is the same. Expressions with radicals cannot be added or subtracted unless *both* the root power *and* the value under the radical are the same.

Here are some examples showing what I mean.

- $\sqrt{2} + \sqrt{3}$

  These *cannot* be combined because it's addition, and the value under the radical is not the same.

- $\sqrt{2} \cdot \sqrt{3} = \sqrt{6}$

These *can* be combined because it's multiplication, and the root power is the same.

✔ $\sqrt{8} \div \sqrt{4} = \sqrt{2}$

These *can* be combined because it's division, and the root power is the same.

✔ $\sqrt{3} - \sqrt[4]{3}$

These *cannot* be combined because it's subtraction, and the root power isn't the same.

✔ $4\sqrt{3} + 2\sqrt{3} = 6\sqrt{3}$

These *can* be combined because the root power and the numbers under the radical are the same.

When the numbers inside the radical are the same, you can see some nice combinations involving addition and subtraction. Multiplication and division can be performed whether they're the same or not. The *root power* refers to square root, $\sqrt{\ }$, cube root, $\sqrt[3]{\ }$, fourth root, $\sqrt[4]{\ }$, and so on.

The rules for adding, subtracting, multiplying, and dividing radical expressions are best summarized below.

Radical Rules: Assume that a and b are positive values.

✔ $m\sqrt{a} + n\sqrt{a} = (m + n)\sqrt{a}$

Addition and subtraction can be performed if the root power and value under the radical are the same.

✔ $m\sqrt{a} - n\sqrt{a} = (m - n)\sqrt{a}$

✔ $\sqrt{a} \cdot \sqrt{a} = \sqrt{a^2} = a$

✔ $\sqrt{a} \cdot \sqrt{b} = \sqrt{ab}$

Multiplication and division can be performed if the root powers are the same.

✔ $\sqrt{a} / \sqrt{b} = \sqrt{a/b}$

Table 4-1 gives you some common square roots.

Table 4-1	Common Square Roots	
$\sqrt{1} = 1$	$\sqrt{4} = 2$	$\sqrt{9} = 3$
$\sqrt{16} = 4$	$\sqrt{25} = 5$	$\sqrt{36} = 6$
$\sqrt{49} = 7$	$\sqrt{64} = 8$	$\sqrt{81} = 9$
$\sqrt{100} = 10$	$\sqrt{10,000} = 100$	$\sqrt{1,000,000} = 1,000$

TIP

Notice that the square root of a 1 followed by an even number of zeros is always a 1 followed by half that many zeros.

The convention that mathematicians have adopted is to use fractions in the powers to indicate that this stands for a root or a radical.

✔ $\sqrt{x} = x^{1/2}$

✔ $\sqrt[3]{x} = x^{1/3}$

✔ $\sqrt[4]{x} = x^{1/4}$

✔ $\sqrt{4ab} = (4ab)^{1/2} = (4)^{1/2} a^{1/2} b^{1/2} = 2a^{1/2} b^{1/2}$

✔ $\sqrt[3]{x^2 y} = (x^2)^{1/3} y^{1/3} = x^{2/3} y^{1/3}$

TIP

When there's no number outside and to the upper left of the radical, you assume that it's a two.

Recall that when raising a power to a power, you multiply the exponents. This is discussed in the "Powers of Powers" section, earlier in the chapter.

ALGEBRA RULES
$\dfrac{1}{+1}$
$\dfrac{}{2}$

When changing from radical form to fractional exponents:

✔ $\sqrt[n]{a} = a^{1/n}$

The *n*th root of *a* can be written as a fractional exponent with *a* raised to the reciprocal of that power.

✔ $\sqrt[n]{a^m} = a^{m/n}$

When the *n*th root of a^m is taken, it's raised to the $\frac{1}{n}$ power. Using the "Powers of Powers" rule, the *m* and the $\frac{1}{n}$ are multiplied together.

This rule allows you to simplify the following expressions. Note that when using the "Powers of Powers" rule the bases still have to be the same.

✔ $6x^2 \cdot \sqrt[3]{x} = 6x^2 \cdot x^{1/3} = 6x^{2 + 1/3} = 6x^{7/3}$

✔ $3\sqrt{x} \cdot \sqrt[4]{x^3} \cdot x = 3x^{1/2} \cdot x^{3/4} \cdot x^1 = 3x^{9/4}$

Leave the exponent as $\frac{9}{4}$. Don't write it as a mixed number.

✔ $4\sqrt{x} \cdot \sqrt[3]{a} = 4x^{1/2} a^{1/3}$

These can't really be combined because the bases are not the same.

The Cheat Sheet repeats the powers rules, so check there if you get confused.

Chapter 5

Doing Operations in Order and Checking Your Answers

Algebra had its start as expressions that were all words. Everything was literally spelled out. As symbols and letters were added, algebraic manipulations became easier. But, as more symbols were added, the rules that went along with the symbols also became a part of the algebra. All this shorthand is wonderful, as long as you know the rules and follow the steps that go along with them. The *order of operations* is a biggie that you use frequently. It tells you what to do first, next, and last in a problem, whether terms are in grouping symbols or raised to a power.

And, because you may not always remember the order of operations correctly, it's very important to check your work. Making sure that the answer you get makes sense, and that it actually solves the problem is the final step of working every problem. The very final step is writing the solution in a way that other folks can understand easily.

This chapter walks you through the order of operations, checking your answers, and writing them correctly.

Ordering Operations

When does it matter in what *order* you do things? Or does it matter at all? Well, take a look at a couple of real-world situations:

> ✔ When you're cleaning the house, it *doesn't* matter whether you clean the kitchen or the living room first.

> ✔ When you're getting dressed, it *does* matter whether you put your shoes on first or your socks on first.

Sometimes the order matters, sometimes it doesn't. It does matter in what order *different* mathematical operations are performed. If you're doing only addition or you're doing only multiplication, you can use any order you want. But, as soon as you mix things up with different operations in the same expression, then you have to pay close attention to the correct order. You can't just pick and choose what to do first, next, and last according to what you feel like doing.

For example, look at the different ways this problem could be done, if there were no rules. Notice that all four operations are represented here.

Solve for the result: $8 - 3 \times 4 + 6 \div 2 =$

One way to do the problem is to just go from left to right:

1. $8 - 3 = 5$

2. $5 \times 4 = 20$

3. $20 + 6 = 26$

4. $26 \div 2 = 13$

Giving a final answer of 13.

Another approach is to group the 3×4 together. Remember, grouped terms tell you that you have to do the operation inside the grouping symbol first.

1. $8 - (3 \times 4) = 8 - 12 = -4$

2. $-4 + 6 = 2$

3. $2 \div 2 = 1$

Giving a final answer of 1.

Using other groupings, you can get answers of 25, 60, or even zero. I won't go into how these answers are obtained because they're all wrong, anyway.

Mathematicians designed rules so that anyone reading a mathematical expression could do it the same way and get the same answer. In the case of multiple signs and operations, working out the problems needs to be done in a specified *order,* from the first to the last. This is the *order of operations.*

Order of operations: Work out the operations and signs in the following order:

1. **Powers or roots**

2. **Multiplication or division**

3. **Addition or subtraction**

Calculating differences in brackets

In algebra problems, parentheses, brackets, and braces are all used for grouping. Terms inside the grouping symbols have to be operated upon before they can be acted upon by anything outside the grouping symbol. All the bracket types have equal weight; none is more powerful or acts differently from the others. This does not carry through with many graphing calculators. The brackets and braces mean something entirely different in those instruments. In most graphing calculators,

✔ **Brackets** mean that the items inside are a part of a matrix, a rectangular arrangement of numbers.

✔ **Braces** mean that what's inside is part of a list of numbers.

These differences make for some awkward situations when you want to show several groupings within a single expression. Because you're limited to parentheses only, and they're all the same size, there's often confusion as to when a grouping starts and where it ends.

When you have two operations on the same "level," you can do those in any order. So, if there's both a power and a root, either can be done first. If you have more than two operations, do them in order from left to right, following the order of operations.

If the problem contains grouped items, do what's inside a grouping symbol first, then follow the order of operations. The grouping symbols are

✔ **Parentheses ():** Parentheses are the most commonly used symbols for grouping.

✔ **Brackets: [ ] and Braces: { }:** Brackets and braces are also used frequently for grouping and have the same effect as parenthesis. Using the different types of symbols helps when there's more than one grouping in a problem. It's easier to tell where a group starts and ends.

✔ **Radical:** $\sqrt{}$: This is used for finding roots.

✔ **Fraction Line (Vinculum)** ————: The fraction line also acts as a grouping symbol — everything above the line in the numerator is grouped together, and everything below the line in the denominator is grouped together.

Even though the order of operations and grouping symbol rules are fairly straightforward, it's hard to describe, in words, all the situations that can come up in these problems. The examples in this chapter should clear up any questions you may have.

The following problem doesn't have parentheses or brackets to indicate what operation needs to be done first, but the order of operations rule (cited earlier in this section) says that multiplication and division need to be worked out before addition and subtraction. To find the solution to $8 - 3 \times 4 + 6 \div 2 =$, for example, follow these steps:

1. Multiply the 3×4 and divide the $6 \div 2$ to get 12 and 3.

   The parentheses help emphasize what to do first.

   $8 - (3 \times 4) + (6 \div 2) = 8 - 12 + 3$

2. Add and subtract in order from left to right.

   $8 - 12 + 3 = -1$

   $8 - 3 \times 4 + 6 \div 2 = -1$

In this example, the operations are grouped to help you.

$$\left[8 \div (5 - 3)\right] \times 5 =$$

1. Subtract the 5 - 3 in the parenthesis to get 2.

2. Divide 8 by 2 and multiply that answer by 5.

   $[8 \div 2] \times 5 =$

3. Multiply 4 times 5.

   $[8 \div 2] \times 5 = 4 \times 5 = 20$

Don't let the division in the next problem put you off. It's easy!

To solve $\dfrac{4(7 + 5)}{2 + 1}$

1. Add the 7 and 5 in the numerator, then the 2 and 1 in the denominator.

   $\dfrac{4(12)}{3}$

   Remember that the fraction line is a grouping symbol. The 2 and 1 in the denominator are grouped together and have to be added first, and before you divide the sum into the numerator.

2. Multiply and divide to get.

   $\dfrac{4(12)}{3} = \dfrac{48}{3} = 16$

Although any operations in parentheses or brackets take precedence, exponents and roots should be solved first, according to the order of operations. Now you can work out a problem with exponents.

$2 + 3^2 (5 - 1) =$

1. Subtract the 1 from the 5 in the parenthesis to get 4.

   $2 + 3^2 (4)$

2. Raise the 3 to the second power to get 9.

$2 + 9(4)$

3. Multiply the 9 and 4 to get 36.

$2 + 9(4) = 2 + 36$

4. Calculate the final answer.

$2 + 36 = 38$

Try this problem involving a square root on for size. Remember to work any operation in parentheses or brackets first.

$$(3 + 4)\sqrt{25} - 8$$

1. Add the 3 + 4.

$7\sqrt{25} - 8$

2. Find the square root of 25. (Check the Cheat Sheet at the front of the book for a list of square roots.)

$7 \cdot 5 - 8$

3. Multiply the 7 and 5.

$35 - 8$

4. Solve the problem.

Since $35 - 8 = 27$, then $(3 + 4)\sqrt{25} - 8 = 27$.

Be sure to catch the subtle difference between the two expressions here: -2^4 *and* $(-2)^4$. The expression $-2^4 = -16$ because the order of operations says to first raise to the fourth power and then apply the minus sign.

The expression $(-2)^4 = 16$ because the entire result in the parenthesis is raised to the fourth power. This is equivalent to multiplying −2 by itself four times. That's an even number of negative signs, so the result is positive.

Engineering the great pyramids

The Egyptians were strong in geometry and developed rough formulas to find the volume of various solid figures. They put this knowledge to work building the pyramids, which are not only engineering marvels but mathematical marvels as well. With relatively primitive tools, such as levers and plumb lines, the Egyptians cut stones that varied, on average, by about $\frac{1}{100}$ of an inch, and they were brought together to within $\frac{1}{500}$ inch. The Great Pyramid of Cheops has sides that face directly north, south, east and west, correct to $\frac{1}{12}$ of a degree.

In general, if you want a negative number raised to a power, you have to put it in parentheses with the power outside.

Checking Your Answers

Checking your answers when doing algebra is always a good idea, just like reconciling your checkbook with your bank statement is a good idea. Actually, checking answers in algebra is easier and more fun than reconciling a checking account. Or, maybe your checking account is more fun than mine.

Check your answers in algebra on two levels.

> ✔ **Level 1:** Does the answer make any sense?
>
> If your checkbook balance shows $40 million, does that make any sense? Sure, we'd all *like* it to be that, but for most of us this would be a red flag that something is wrong with our computations.

> ✔ **Level 2:** Does actually putting the answer back into the problem give you a true statement? Does it *work?*
>
> This is the more critical check because it gives you more exact information about your answer. The first level helps weed out the obvious errors. This is the final check.

The next sections help you make even more sense of these checks.

Making sense at level 1

To check as to whether an answer makes any sense or not, you have to know something about the topic. These situations should involve things you're familiar with. Just use your common sense.

For instance, your answer to an algebra problem is $x = 5$. If you're solving for Jon's weight in pounds, unless Jon is a guinea pig instead of a person, you probably want to go back and redo the work. Five pounds doesn't make any sense as an answer in this context.

On the other hand, if the problem involves a number of pennies in a person's pocket, then five pennies seems reasonable. Getting five as the number of home runs a player hit in one ballgame may at first seem quite possible, but if you think about it, five home runs in one game is a lot — even for Sammy Sosa or Mark McGuire. You may want to double-check.

Plugging in level 2

Actually plugging in your answer requires you to go through the algebra manipulations in the problem. You add, subtract, multiply, and divide to see if you get a true statement using your answer.

For example, suppose Jack has four more pennies than Jill. If they have a total of 14 pennies altogether, then how many pennies does Jill have? Does $x = 5$ work for an answer?

1. **Write the problem.**

   Let x represent the number of pennies that Jill has. Jack has $x + 4$ pennies. That means, $x + (x + 4) = 14$.

   The number of pennies Jill has plus the number of pennies Jack has equals 14.

2. **Insert the answer into the equation.**

   Replace the variable with 5 to get $5 + (5 + 4) = 14$.

3. **Do the operations and check to see if the answer works.**

   $5 + 9 = 14$ is a true statement, so the problem checks. Jill has 5 pennies; Jack has 4 more than that, or 9 pennies; altogether, they have 14 pennies.

You can apply a variation of the preceding steps to check whether $x = 2$ works in the following equation

$$5x\,[x + 3\,(x^2 - 3)] + 1 = 0$$

Replace the variable with 2 to get $5 \cdot 2\,[2 + 3\,(2^2 - 3)] + 1 = 0$

Do the operations and simplify:

Square the 2 to get $5 \cdot 2\,[2 + 3\,(4 - 3)] + 1 = 0$

Subtract in the parenthesis to get $5 \cdot 2\,[2 + 3\,(1)] + 1 = 0$

Add in the bracket to get $5 \cdot 2\,[5] + 1 = 0$

Multiply the 5, 2 and 5 to get $50 + 1 \neq 0$

This time the work does *not* check. You should go back and try again to find a value for x that works.

Writing Understandable Answers

An algebraic expression may be written in all sorts of ways, which may all be *correct,* but the different ways of writing may not necessarily be nice, pretty, or useful. Yes, algebra can be pretty, and in this section I show you how.

Look at three versions of the same answer to a question. Which do you prefer?

A) $3adce - becd4 + aed5b - 2$

B) $3acde - 4bcde + 5abde - 2$

C) $ed3ac - b4edc - 2 + 5bade$

Which is your choice? Does one answer look neater and easier to read? (I'm going with choice B.) Following certain mathematical conventions makes your notation uniform, understandable, and easier to compare to other algebraic expressions.

In any one term, put the numbers first, and then let the variables follow in alphabetical order. Put radicals at the end.

To put $ed3ac$ in standard format, write the number first, then alphabetize the variables: $3acde$.

When arranging a string of terms, take all the terms containing the same variable and put them in either increasing or decreasing powers of that variable. For example, to put terms in decreasing powers of a variable, find the term that contains the highest power of that variable, put it first, and then find the next highest, put it second, and so on. The choice of variable and whether it's increasing or decreasing depends on the situation and what you're going to do next. If a problem involves solving for the value of the variable x, and the problem has several powers of x, write the terms in order of the powers of x — whether the order is decreasing or increasing.

To illustrate this, check out the following examples. Each individual term is written correctly, but the terms need to be put in some order.

✔ Option 1 is in *increasing* powers of x. Note that the lowest power of x comes first.

$$3xy^2 + 8x^2y^3 - 4x^3y$$

✔ The second option is in *decreasing* powers of x. This time the highest power of x comes first.

$$-4x^3y + 8x^2y^3 + 3xy^2$$

✔ You can switch variables, and write in terms of the *increasing* powers of y.

$$-4x^3y + 3xy^2 + 8x^2y^3$$

✔ Stick with y, but write the terms in *decreasing* powers of y.

$$8x^2y^3 + 3xy^2 - 4x^3y$$

Using these conventions makes your notation more understandable to the people who read your work.

Chapter 6

Prepping for Operations

*W*ork in algebra is done using variables (letters) and symbols (see more on variables and symbols in Chapter 1). These are like shorthand notation — quicker and easier to write than all the big words in longhand. But, with shorthand, you have to know what everything means and how it goes together. In algebra, simple arithmetic rules are transformed into rules for adding, subtracting, multiplying, and dividing variables. Letting variables represent numbers allows for more flexibility, but you have to be careful — everything is disguised.

Realizing Some Restrictions

When algebra uses variables to represent numbers that can be added, subtracted, multiplied, and divided, you assume that the variables are representing *quantities* or *amounts* that can be added, subtracted, multiplied, and divided. But using the representation is not quite that simple or obvious. Even when you're just adding numbers together, restrictions exist. Likewise, there are restrictions and rules when you're adding variables together or numbers and variables together.

Wait a minute! What *restrictions* are there for just adding *numbers* together? Why would there be any problem with that? Doesn't one plus one still equal two? Sure, unless it's equal to four. Seriously, consider what happens when you add six quarters and four dimes. When you add the *numbers* together, what do you get (aside from not enough money for a gourmet cup of coffee)? Ten quarters? Ten dimes? Ten duarters? Ten quimes?

No. This is silly, of course. But it illustrates what I mean by restrictions on adding numbers together. If you want to add quarters and dimes, then count the number of coins and say that you have ten *coins,* or change to the money value of each coin and say that you have $1.90. When adding quantities or amounts, you have to be sure that the amounts *can* be added. That's even more critical when you add letters because silly errors aren't as obvious. Care has to be taken to add them correctly.

Representing Numbers with Letters

"This segment is sponsored by the letter *b,* standing in for a number to be determined later." Okay, so that's a bad take-off on *Sesame Street,* but I hope you see what I'm driving at. In algebra, letters stand in for numbers all the time. Just keep in mind that letters, or variables, *always* stand for numbers.

Letting a variable represent a quantity can simplify a problem and lead to nice, neat situations because you don't have to deal with a bunch of messy words. Sometimes you do have to deal with some fairly complicated situations. But fear not! A few simple rules can help change even the most complicated situation into an easily understandable one.

Attaching factors and coefficients

One nice thing about algebra is that it conserves energy — the energy that would be needed to write multiplication symbols between letters. Even having to write dots between symbols takes time, so a simpler system was devised. When a number is written in front of a variable, such as $3x$, it means that the 3 and x are multiplied together. The 3 is a *coefficient* in this case.

A number preceding a variable is a *coefficient.* For example, the number four is the coefficient when $a + a + a + a$ is expressed as $4a$.

When several variables are multiplied together, multiplication symbols aren't needed. The term $3xyz$ means that all four factors are multiplied together.

Provided that the number is directly in front of the variable, $2a$ means that 2 and a are multiplied together. So $2a$ is two times, or twice, a. Twice as many apples means that if a represents 10 apples, then $2a$ represents 2 times 10, or 20 apples.

Boolean algebra

The use of symbols for statements and complex relationships isn't limited to algebra. George Boole, an Englishman, developed symbols to deal with difficult logic problems. His system is called Boolean algebra. An upside-down v, ∧, represents the word *and*. A right-side-up v, ∨, represents the word *or*. The letters P, Q, and R represent statements such as P: It is raining; Q: The sun is out. When a statement such as P ∧ Q is given (It is raining and the sun is out.), then it can be judged as true or false depending on whether the individual statements are true. A method of truth tables is used in logic to test the validity of arguments based on the statements. Boole's system is still the basis for the study of logic today.

For example, if *a* represents the number of apples you have, what does 2*a* represent?

- ✔ Two *more* apples?
- ✔ *Twice as many* apples?
- ✔ *Half* as many apples?

The answer is: 2*a* represents twice the number of apples that you have.

The symbols + and − may mean many things to you. They mean many things in algebra, too. It all depends on the context. These symbols always separate *terms,* which are clusters of variables and numbers connected by multiplication and division. Plus and minus signs separate terms from one another.

In algebra-speak, a plus sign means *and, more, increased by, added to,* and so on. The expression 2 + *a* could represent

- ✔ A gift of *a* dollars has increased my pocket change up from two dollars.
- ✔ Two people went through the door, and then *a* more went in.
- ✔ The temperature was two degrees, and then it went up *a* degrees.

This is all different from the term 2*a* in which the coefficient 2 doubles the amount of the variable *a*.

The minus sign means *less, take away, decreased by, subtracted from,* and so on. The expression *a* − 2 could represent:

> ✔ There were a administrators, but their number was decreased by two.
>
> ✔ There were two less than a alligators in the pond.
>
> ✔ William Tell had a apples when he started and two fewer when he finished.

Even though you have to take care when letting variables represent numbers, the benefit and ease in working with variables outweighs the possible difficulties. Besides the advantage of not having to write as much, it's easier to focus on a problem that takes up less space — your eyes can track better. Also, algebraic symbols are precise. The hidden meanings written words can have don't exist in algebra — a universal language that crosses the boundaries that language and time can present. The beauty of variables can be appreciated in the following exercise:

Consider the task of collecting, organizing, and reporting on the coins collected during a charity drive.

Let n represent the value of the nickels in a roll of nickels.

Let d represent the value of the dimes in a roll of dimes.

Let q represent the value of the quarters in a roll of quarters.

After collecting the money, you get the following information from your helpers:

Ann collected six rolls of nickels, four rolls of dimes, and nine rolls of quarters, or $6n + 4d + 9q$.

Ben collected five rolls of nickels, three rolls of dimes, and seven rolls of quarters or $5n + 3d + 7q$.

Cal collected 15 rolls of nickels, two rolls of dimes, and six rolls of quarters, or $15n + 2d + 6q$.

Don collected one roll of nickels, three rolls of dimes and four rolls of quarters, or $n + 3d + 4q$.

Set up an equation to add all the rolls of nickels, dimes, and quarters:

$$(6 + 5 + 15 + 1)n + (4 + 3 + 2 + 3)d + (9 + 7 + 6 + 4)q =$$

$$27n + 12d + 26q$$

The total amount of money can now be computed if you know that

A roll of nickels is worth \$2, so $n = 2$.

A roll of dimes is worth \$5, so $d = 5$.

A roll of quarters is worth \$10, so $q = 10$.

So, $27n + 12d + 26q = 27(2) + 12(5) + 26(10) = 374$

A total of $374 was raised.

If you just deal with the number of rolls of each coin and save the money total for last, you keep the computation much simpler. The numbers don't have to be as large.

Doing the Math

Addition was probably the first operation you discovered. It's the easiest for people of all ages to picture and relate to. Adding is a bit trickier in algebra just because so often you *can't* add. But when you can, it's a nice process.

Adding and subtracting variables

When adding, instead of expressing $a + a + a + a$ the long way, you can just write $4a$, which says the same thing more efficiently because multiplication is just repeated addition. In the case of $4a$, the number represented by a is added four times. Or, you can say that a is multiplied by four.

By expressing $a + a + a + a$ as $4a$, however, you are creating a single *term,* which is math jargon for coefficient(s) and/or variable(s) grouped together but sometimes separated from other terms by a plus or minus sign. So, the following operation, where the variables $a, b,$ and c represent any real numbers, has three *terms:*

$10ab + 3c - 7c$

When adding or subtracting terms that have *exactly* the same variables, combine the coefficients.

When adding $2a + 5a + 4a$ what is the result?

Because you have three separate quantities, and each of them has an a, you can add them together as long as the a represents the same thing in each one. One a can't represent the number of apples while the other a represents the number of aardvarks. They all have to represent the same thing in the same problem.

A variable that appears more than once in an expression or equation always represents the same number. If the variable could represent more than one thing, the statement would be worthless — with no way to tell one meaning of the variable from another.

It's nice when the variable chosen to represent some number can start with the same letter as what it represents, such as a for aardvark. But this isn't necessary. A letter/name coordination is useful when a problem is composed of more than one variable, but taking careful notes and identifying variables works just as well.

Now try adding up the same variable in the following example:

$2a = a + a$

$5a = a + a + a + a + a$

$4a = a + a + a + a$

That's a total of 11 a variables altogether. Notice that the numbers in front, the coefficients 2, 5, and 4, add up to 11, also.

$2a + 5a + 4a = (2 + 5 + 4)a = 11a$

When adding or subtracting terms with the same variable (such as x or n in the following examples), add or subtract the coefficients (numbers in front of the variables), and let the result stand alongside the variable. If a, b, and c are coefficients of the variable x, then

$ax + bx - cx = (a + b - c)x.$

The following examples further demonstrate how to add terms with the same variable:

$3x + 2x = 5x$

$9n + 6n + n = 16n$

When there is no number in front of the variable, assume that the number is one. (This is one of the few times you can assume something and not make a donkey of yourself.)

$a = 1a$

$x = 1x$

The following example shows you how one variable can be added to another term with the same variable.

$a + 3a + x + 2x = 1a + 3a + 1x + 2x = 4a + 3x$

Notice that you add terms that have the same variables because they represent the same amounts. You don't try to add the terms with different variables. The examples that follow involve two or more variables:

$5a + 2a + 6b + 8b + 11c = 7a + 14b + 11c$

$3x + 4y - 2x - 8y + x = 2x - 4y$

When subtracting terms, use the rules for adding and subtracting signed numbers and apply them to the coefficients. (Check out Chapter 2 for information on working with signed numbers.)

$$5a + 4a - 2a + 6 + 3b - 2b = 7a + b + 6$$

Notice that the 6 doesn't have a variable. It stands by itself; it isn't multiplying anything.

Adding and subtracting with powers

The following examples show how addition and subtraction are performed on several terms involving variables. Notice that the terms that combine *always* have exactly the same variables with exactly the same powers. For more on powers (exponents), see Chapter 4.

- ✔ $x + x + x = 3x$
- ✔ $x^2 - 2x^2 + 3x^2 + 3x^2 = 5x^2$
- ✔ $x + 3x + 4x^2 + 5x^2 + 6x^3 = 4x + 9x^2 + 6x^3$
- ✔ $4x^4 - 3x^3 + 2x^2 + x - 1$

In the last example, none of the powers (exponents) are the same, so even though the variables are the same, you can't add the numbers in front together.

In order to add or subtract terms with the same variable, the exponents of the variable must be the same. Perform the required operations on the coefficients, leaving the variable and exponent as they are.

Because x and x^2 don't represent the same amount, they can't be added together.

$$3a^3 + 3a^2 + 3a + a + 2a^2 + 2a^4 = 2a^4 + 3a^3 + (3 + 2)a^2 + (3 + 1)a =$$
$$2a^4 + 3a^3 + 5a^2 + 4a$$

Notice that the exponents are listed in order from highest to lowest. This is a common practice to make answers easy to compare.

$$2m + 3m^2 + 5m^3 - 2m^2 - 3m - 1 = 5m^3 + (3 - 2)m^2 + (-3 + 2)m - 1 =$$
$$5m^3 + m^2 - m - 1$$

An expression like $5x + 3x^2 + 2y - 4y^2 + 3$ cannot be simplified. There are terms with the same exponent ($3x^2$ and $4y^2$) but the variables (x and y) aren't the same.

Multiplying and Dividing Variables

Multiplying variables is in some ways easier than adding or subtracting them, but you still have to be a bit careful. When you divide variables, however, you need to follow some relatively strict rules. In the following sections, I give you the tips and rules.

Multiplying variables

When the variables are the same, multiplying them together "compresses" them into a single factor (variable). But you still can't combine *different* variables.

When multiplying variables, multiply the coefficients and variables as usual. If the bases are the same, you can multiply the bases by merely adding their exponents. (See more on the multiplication of exponents in Chapter 4.) Write the result in a compact form.

Look at these examples where the letters are the variables as well as the bases.

> ✔ $a \cdot a \cdot b \cdot c = a^2 bc$
>
> ✔ $2 \cdot a \cdot a \cdot a \cdot b \cdot b \cdot c = 2a^3 b^2 c$
>
> ✔ $2 \cdot a \cdot a \cdot a \cdot a \cdot 3 \cdot b \cdot b \cdot b \cdot 4 \cdot c \cdot c = 24a^4 b^3 c^2$
>
> ✔ $2 \cdot a^2 \cdot a^3 \cdot 3 \cdot b \cdot b \cdot b^6 \cdot 5 \cdot c \cdot c^2 \cdot c^{10} = 30a^5 b^8 c^{13}$
>
> ✔ $(2a^2 b^2 c^3)(4a^3 b^2 c^4) = 8a^{2+3} b^{2+2} c^{3+4} = 8a^5 b^4 c^7$
>
> ✔ $(3x^2 yz^{-2})(4x^{-2} y^2 z^4)(3xyz) = 36x^{2-2+1} y^{1+2+1} z^{-2+4+1} = 36x^1 y^4 z^3$

Dividing variables

When you want to divide a combination of variables and numbers, divide the numbers as if you're reducing fractions (see Chapter 3 for fraction reduction). But only variables that are alike can be divided.

In division, the answers don't have to come out even. There can be a remainder — a value left over when one number is divided by another. But you don't want remainders here (they'd be new terms). So, be sure you don't leave any remainders lying around.

When dividing variables, write the problem as a fraction. Using the greatest common factor, divide the numbers and reduce. Use the rules of exponents (see Chapter 4 or the Cheat Sheet) to divide variables that are the same.

Fraction Line

Solidus or *virgule,* two different words, denote the slanted line (/) used for fractions. The solidus is also found in nonmathematical settings, such as and/or and miles/hour. In poetry, the solidus shows line breaks as in "Old King Cole/Was a merry old soul." When the fraction line is horizontal and has values above and below it, it's called a *vinculum.*

Dividing variables is fairly straightforward. Each variable is considered separately. The number coefficients are reduced the same as in simple fractions.

This can be explained with aluminum cans: Four friends decided to collect aluminum cans for recycling (and money). They collected $12x^3$ cans, and they're going to get y^2 cents per can. The total amount of money collected is then $12x^3 y^2$ cents. How will they divvy this up?

Divide the total amount by four to get the individual amount that each of the four friends will receive. $\frac{12x^3 y^2}{4} = 3x^3 y^2$ cents each. The only thing that divides here is the coefficient.

If you want the number of cans each will get paid for, divide $\frac{12x^3 y^2}{4y^2} = 3x^3$ cans.

Why is using variables better than using just numbers? Because if the numbers change, then you still have all the shares worked out. Just let the x and y change in value. Work through the following examples to find out how to divide using variables, coefficients, and exponents:

✔ If a stands for the number of apples, ten apples divided into groups of five apples each results in two groups (not two apples).

$$\frac{10a}{5a} = 2a$$

Ten apples divided into five groups results in two apples per group.

✔ $\frac{6a^2}{3a} = 2a$

Three divides six twice. Using the rules of exponents, $a^2 \div a = a$.

I prefer to write the answer with x in the denominator and a positive exponent rather than in the numerator with a negative exponent.

✔ $\frac{14x^2}{7x^4} = 2x^{-2} = \frac{2}{x^2}$

✔ $\frac{6x^3 y^2}{18xy^4} = \frac{x^2}{3y^2}$

Doing It All

The four main operations, addition, subtraction, multiplication, and division, are covered in the preceding sections. Many algebra problems involve more than one operation, so look at the following steps to see how to handle a combination of operations.

In this example, the operations are performed on

$$4a^2 b^3 (2a^3 b^2) + 5ab^{-2} (2a^4 b^7) + 5$$

1. **Multiply the variables together separately in each term.**

   $$4 \cdot 2a^2 a^3 b^3 b^2 + 5 \cdot 2aa^4 b^{-2} b^7 + 5$$

2. **Add the exponents of the variables that are alike.**

   $$8a^{2+3} b^{3+2} + 10a^{1+4} b^{-2+7} + 5 =$$
   $$8a^5 b^5 + 10a^5 b^5 + 5$$

   You can see that the first two terms are alike as far as the variables they have and the exponents on those variables, which is why you can add them together.

3. **Combine terms that are alike.**

   $$= (8 + 10) a^5 b^5 + 5 = 18a^5 b^5 + 5$$

Okay. Now that you have successfully met the challenge of performing several operations on one complex example, why not try going through the steps again to perform a combination of operations on another example?

$$3m^2 (2mn) - 4m^3 n^3 (2n^{-2}) + 5m^2 n^3 - 6mn (mn)$$

1. **Multiply the variables together separately in each term.**

   $$3 \cdot 2m^2 mn - 4 \cdot 2m^3 n^3 n^{-2} + 5m^2 n^3 - 6mmnn$$

2. **Add the exponents of the variables that are alike.**

   $$6m^{2+1} n - 8m^3 n^{3-2} + 5m^2 n^3 - 6m^{1+1} n^{1+1} =$$
   $$6m^3 n - 8m^3 n + 5m^2 n^3 - 6m^2 n^2$$

3. **Combine the terms that are alike.**

   In this case, only the first two terms can be combined; their variables and their exponents match.

   $$(6 - 8) m^3 n + 5m^2 n^3 - 6m^2 n^2 =$$
   $$-2m^3 n + 5m^2 n^3 - 6m^2 n^2$$

The following example is your chance to strut your stuff. You've done the multiplying, so the next step is division (which is really simple subtraction). Go for it!

In this example, the operations are performed on

$$\frac{4x^2 y^3}{2xy} - \frac{15xy^5}{3y^3} + \frac{13x^{-2} y^{11}}{x^{-5} y^8} + \frac{11x^4 y^{7/2}}{xy^{1/2}}$$

1. **Divide by subtracting the exponents of the common bases.**

   Divide the known numbers. Assume that the base without an exponent has one for an exponent. This problem has negative exponents to deal with in both the numerator and the denominator.

   $$2x^{2-1} y^{3-1} - 5xy^{5-3} + 13x^{-2+5} y^{11-8} + 11x^{4-1} y^{7/2-1/2}$$

2. **Complete the subtraction on the exponents.**

   Note: When the negative exponent (–5) that was in the denominator was brought up, it became positive and was added. Fractional exponents work just like other whole number exponents; they add and subtract just the same.

   $$= 2xy^2 - 5xy^2 + 13x^3 y^3 + 11x^3 y^3$$

3. **Add or subtract the terms that are exactly alike — numbers that have variables and exponents in common.**

   $$(2-5)xy^2 + (13+11)x^3 y^3 =$$
   $$-3xy^2 + 24x^3 y^3$$

In this last example, the rule for multiplying exponents is used in two of the terms. For more on this, see the Cheat Sheet and Chapter 4.

Perform the operations on $(3x^2 y)^2 + 5x^4 y - (2xy^{1/2})^4 - 16xy^4$

1. **Raise powers to powers by multiplying the exponents.**

   $$3^{1 \cdot 2} x^{2 \cdot 2} y^{1 \cdot 2} + 5x^4 y - 2^{1 \cdot 4} x^{1 \cdot 4} y^{(1/2) \cdot 4} - 16xy^4$$

2. **Multiply the values in the exponents.**

   $$3^2 x^4 y^2 + 5x^4 y - 2^4 x^4 y^2 - 16xy^4 =$$
   $$9x^4 y^2 + 5x^4 y - 16x^4 y^2 - 16xy^4$$

3. **Combine any terms that have variables that are alike.**

   $$(9-16)x^4 y^2 + 5x^4 y - 16xy^4 =$$
   $$-7x^4 y^2 + 5x^4 y - 16xy^4$$

You deal with terms that have exponents outside the parenthesis first. Then you can combine the terms that have the exact same powers on x and y.

Part II

Figuring Out Factoring

In this part . . .

In these chapters, you find out how to rearrange algebraic expressions to make them more usable. Factoring is always a high priority and is covered in depth, as well as familiar and not-so-familiar ways to manage your algebraic expressions, including the FOIL happy face in Chapter 10. Mastering the methods here makes the rest possible.

Chapter 7

Working with Numbers in Their Prime

*P*rime numbers (whole numbers evenly divisible only by themselves and one) have been the subject of discussions between mathematicians and nonmathematicians for centuries. Prime numbers and their mysteries have intrigued philosophers, engineers, and astronomers. These folks and others have discovered plenty of information about prime numbers, but many unproven conjectures remain.

Probably the biggest mystery is determining what prime number will be discovered next. Computers have aided the search for a comprehensive list of prime numbers, but because numbers go on forever without end, and because no one has yet found a pattern or method for listing prime numbers, the question involving the *next one* remains.

Beginning with the Basics

Prime numbers are important in algebra because they help you work with the smallest possible numbers. Big numbers are often unwieldy and can produce more computation errors. So reducing fractions to their lowest terms and factoring expressions to make problems more manageable are basic tasks.

A *prime number* is a whole number larger than the number one that can be divided evenly only by itself and one.

The first and smallest prime number is the number two. It's the only *even* prime number. All primes after two are odd because all even numbers can be divided evenly by one, themselves, and two. They don't fit the definition of a prime number.

If a number divides evenly by 3, adding up the digits in the number gives you a multiple of 3. Check out Chapter 22 for more divisibility rules.

Table 7-1 lists the first 20 prime numbers. Confirm for yourself that each can be divided evenly *only* by itself and one.

Table 7-1		Prime Numbers under 100		
2	3	5	7	11
13	17	19	23	29
31	37	41	43	47
53	59	61	67	71
73	79	83	89	97

When you recognize that a number is prime, you don't waste time trying to find things to divide into it when you're reducing a fraction or factoring an expression. There are so many primes that you can't memorize or recognize them all, but just knowing or memorizing the primes smaller than 100 is a big help. See the Cheat Sheet at the front of the book for a list of the first 100 prime numbers.

Why isn't the number 1 prime?

By tradition and definition, the number 1 is not prime. The definition of a prime number is that it can be divided evenly only by itself and 1. In this case, there would be a double hit, because 1 *is* itself.

Many theorems and conjectures involving primes don't work if 1 is included. Mathematicians around the time of Pythagoras sometimes even excluded the number 2 from the list of primes because they didn't consider 1 or 2 to be *true numbers* — they were just generators of all other even and odd numbers. Sometimes it seems that some rules are a bit arbitrary. In this case, it just makes everything else work better if 1 isn't a prime.

Mersenne Primes

Mersenne Primes are special prime numbers that can be written as one less than a power of two. For example, $2^2 - 1 = 3$; $2^3 - 1 = 7$; and $2^4 - 1 = 15$. Three and 7 are prime numbers, but 15 isn't a prime. So this formula doesn't always give you a prime, it's just that there are many primes that can be written this way.

In 1996, the Great Internet Mersenne Prime Search was launched. This involved a contest to find large Mersenne Primes. A gentleman, on his home computer, found a Mersenne Prime of sufficient size, and the Electronic Frontier Foundation awarded him $50,000. This foundation is offering $100,000 to the first person to discover a 10-million-digit number that is a Mersenne Prime. If you're interested, go to this site on the Internet: www.mersenne.org.

Composing Composite Numbers

Prime numbers are interesting to think about, but they can also be a dead end in terms of factoring algebraic expressions or reducing fractions. The opposite of prime numbers, composite numbers, can be broken down into factorable, reducible pieces. In this section, you'll see how.

Whole numbers larger than one that aren't prime are *composite numbers* that can be broken down into the prime numbers that multiply together to give you that composite number. So, you can write every composite number as the product of prime numbers, a process known as *prime factorization*. Every number's prime factorization is unique.

Some examples of prime factorizations of composite numbers:

- $6 = 2 \cdot 3$
- $12 = 2 \cdot 2 \cdot 3 = 2^2 \cdot 3$
- $16 = 2 \cdot 2 \cdot 2 \cdot 2 = 2^4$
- $250 = 2 \cdot 5 \cdot 5 \cdot 5 = 2 \cdot 5^3$
- $510,510 = 2 \cdot 3 \cdot 5 \cdot 7 \cdot 11 \cdot 13 \cdot 17$
- $42,059 = 137 \cdot 307$

Okay, so the last one is a doozy. Finding that prime factorization without a calculator, computer, or list of primes is difficult. The factors of some numbers aren't always obvious.

Writing Prime Factorizations

Writing the prime factorization of a composite number is one way to be absolutely sure you've left no stone unturned. These factorizations show you the one and only way a number can be factored.

A slick way of writing out prime factorizations is to do an upside-down division. You put a *prime factor* (a prime number that evenly divides the number you're working on) on the outside left and the result or *quotient* (the number of times it divides evenly) underneath. You divide the quotient (the number underneath) by another prime number and keep doing this until the bottom number is a prime. Then you can stop. The order you do this in doesn't matter. You get the same result or list of prime factors no matter what order you use.

Look at the prime factorization of 120.

$$2\overline{)120}$$
$$2\overline{)60}$$
$$2\overline{)30}$$
$$3\overline{)15}$$
$$5$$

Look at the numbers going down the left side of the work and the number at the bottom. They act the same as the divisors in a division problem. Only, in this case, they're all prime numbers. Although many composite numbers could have played the role of divisor for the number 120, the numbers for the prime factorization of 120 must be prime-number divisors.

When using this process, you usually do all the 2s first, then all the 3s, then all the 5s, and so on to make the prime factorization process easier, but you can do this in any order: $120 = 2 \cdot 2 \cdot 2 \cdot 3 \cdot 5 = 2^3 \cdot 3 \cdot 5$. In the next example, start with 13 because it seems obvious that it's a factor. The rest are all in a mixed-up order.

The prime factorization of 13,000:

$$13\overline{)13,000}$$
$$5\overline{)1000}$$
$$2\overline{)200}$$
$$2\overline{)100}$$

$$5\underline{\smash{\big|50}}$$
$$2\underline{\smash{\big|10}}$$
$$5$$

So $13,000 = 13 \cdot 5 \cdot 2 \cdot 2 \cdot 5 \cdot 2 \cdot 5 = 2^3 \cdot 5^3 \cdot 13$

Getting Down to the Prime Factor

Doing the actual factoring in algebra is easier when you can recognize which numbers are composite and which are prime. If you know what they are, then you know what to do with them. Now, try putting all this knowledge to work!

Taking primes into account

Use prime factorization to reduce fractions. Start with numbers only and then add variables (letters that represent any real number) to the mix.

Reduce the fraction $\frac{120}{165}$ by following these steps:

1. **Find the prime factorization of the numerator.**

    120 is $2^3 \cdot 3 \cdot 5$

2. **Find the prime factorization of the denominator.**

    165 is $3 \cdot 5 \cdot 11$

3. **Next, write the fraction with the prime factorizations in it.**

    $$\frac{120}{165} = \frac{2^3 \cdot 3 \cdot 5}{3 \cdot 5 \cdot 11}$$

4. **Cross out the factors the numerator shares with the denominator to see what's left — the reduced form.**

    $$\frac{120}{165} = \frac{2^3 \cdot 3 \cdot 5}{3 \cdot 5 \cdot 11} = \frac{2^3 \cdot \cancel{3} \cdot \cancel{5}}{\cancel{3} \cdot \cancel{5} \cdot 11} = \frac{2^3}{11} = \frac{8}{11}$$

Now, try reducing the fraction $\frac{100}{243}$.

1. Find the prime factorization of the numerator.

    100 is $2^2 \cdot 5^2$

2. Find the prime factorization of the denominator.

    243 is 3^5

3. Write the fraction with the prime factorizations.

$$\frac{100}{243} = \frac{2^2 \cdot 5^2}{3^5}$$

Look at the prime factorizations. You can see that the numerator and denominator have absolutely nothing in common. The fraction can't be reduced. The two numbers are *relatively prime*. The beauty of using the prime factorization is that you can be sure that the fraction's reduction possibilities are exhausted — you haven't missed anything. You can leave the fraction in this factored form or go back to the simpler $\frac{100}{243}$. It depends on your preference.

Reduce the fraction $\frac{48x^3 y^2 z}{84xy^2 z^3}$

1. Find the prime factorization of the numerator.

$$48x^3 y^2 z = 2^4 \cdot 3 \cdot x^3 y^2 z$$

2. Find the prime factorization of the denominator.

$$84xy^2 z^3 = 2^2 \cdot 3 \cdot 7 \cdot xy^2 z^3$$

3. Write the fraction with the prime factorization.

$$\frac{48x^3 y^2 z}{84xy^2 z^3} = \frac{2^4 \cdot 3 \cdot x^3 y^2 z}{2^2 \cdot 3 \cdot 7 \cdot xy^2 z^3}$$

4. Cross out the factors in common.

$$\frac{2^{\cancel{4}2} \cdot \cancel{3} \cdot x^{\cancel{3}2} \cancel{y^2} \cancel{z}}{\cancel{2^2} \cdot \cancel{3} \cdot 7 \cdot \cancel{x} \cancel{y^2} z^{\cancel{3}2}} = \frac{4x^2}{7z^2}$$

By writing the prime factorizations, you can be certain that you haven't missed any factors that the numerator and denominator may have in common.

Pulling out factors — leaving the rest

Pulling out common factors from lists of terms or the sums or differences of a bunch of terms is done for a good reason. It's a common task when you're simplifying expressions and solving equations. The common factor that makes the biggest difference in these problems is the GCF, or greatest common factor. When you recognize the GCF and factor it out, it does the most good.

The *greatest common factor (GCF)* is the largest possible term that evenly divides each term of an expression containing two or more terms.

In any factoring discussions, the GCF, the most common and easiest factoring method, always comes up first. And it's helpful when solving equations. In an expression with two or more terms, finding the greatest common factor can make the expression more understandable and manageable.

The best case scenario is to recognize and pull out the GCF from a list of terms. Sometimes, though, the GCF may not be so recognizable. It may have some strange factors, such as 7, 13, or 23. It isn't the end of the world if you don't recognize one of these numbers as being a multiplier; it's just nicer if you do.

The three terms in the expression $12x^2 y^4 + 16xy^3 - 20x^3 y^2$ have common factors. What is the GCF? These steps help you find it.

1. **Determine any common numerical factors.**

   Each term has a coefficient that is divisible by a power of two, which is four.

2. **Determine any common variable factors.**

   Each term has x and y factors. The prime factorizations should help to show what the GCF is.

3. **Write the prime factorizations of each term.**

   $$12x^2 y^4 = 2^2 \cdot 3 \cdot x^2 y^4$$
   $$16xy^3 = 2^4 \cdot xy^3$$
   $$-20x^3 y^2 = -2^2 \cdot 5 \cdot x^3 y^2$$

   The GCF is the product of all the factors that all three terms have in common.

4. **Find the GCF.**

   The GCF contains the *lowest* power of each variable and number that occurs in any of the terms. Each variable in the sample problem has a factor of 2. If the lowest power of 2 that shows in any of the factors is 2^2, then 2^2 is part of the GCF.

   Each factor has a power of x. If the lowest power of x that shows up in any of the factors is 1, then x^1 is part of the GCF.

   Each factor has a power of y. If the lowest power of y that shows in any of the factors is 2, then y^2 is part of the GCF.

   The GCF is $2^2 xy^2 = 4xy^2$.

TIP

When finding the greatest common factor (GCF) of terms, the lowest power (exponent) of a particular factor that occurs in any of the terms determines the power of that factor in the GCF.

The GCF of $12x^2 y^4 + 16xy^3 - 20x^3 y^2$ is $4xy^2$.

5. **Divide each term by the GCF.**

   The respective terms are divided as shown:

   $$\frac{12x^2 y^4}{4xy^2} = 3xy^2$$

   $$\frac{16xy^3}{4xy^2} = 4y$$

   $$\frac{-20x^3 y^2}{4xy^2} = -5x^2$$

Notice that *all three* results of the division have nothing in common. The first two terms each have a y and the first and third each have an x, but nothing is shared by all the results. This is the best factoring result, which is what you want.

Rewrite the original expression with the GCF factored out and in parentheses:

$$12x^2 y^4 + 16xy^3 - 20x^3 y^2 = 4xy^2 (3xy^2 + 4y - 5x^2)$$

These final examples show you a couple more variations.

- Factor out the GCF of $40a^5 x + 80a^5 y - 120a^5 z$

  The GCF is $40a^5$, so you can write the expression as

  $40a^5 (x + 2y - 3z)$.

- Factor the GCF out of $18x^2 y + 25z^3 + 49z^2$

  Even though none of these terms are prime, the three terms have nothing in common — nothing that *all three* shares. The following prime factorizations demonstrate.

  $$18x^2 y = 2 \cdot 3^2 x^2 y$$

  $$25z^3 = 5^2 z^3$$

  $$49z^2 = 7^2 z^2$$

The last two terms do have a factor of z in common, but the first term doesn't. This expression is said to be *prime* because it cannot be factored.

Chapter 8

Sharing the Fun: Distributing

Algebra is full of contradictory actions. First you're asked to factor (see Chapter 9 for facts on factoring), and then *distribute* or "unfactor." In other words, first you're asked to reduce fractions, and then you're supposed to multiply and create bigger numbers. First you're asked to change a fraction to a decimal and then a decimal to a fraction.

But rest assured that good reasons are behind doing all these seemingly contradictory processes. You carefully wrap a birthday gift so it can be unwrapped the next day. You water and fertilize your lawn to make it grow — just so you can cut it. See, contradictions are everywhere! In this chapter, I tell you when, why, and how to factor and "unfactor." You want to make informed decisions and then have the skills to execute them correctly. What good does it do you to buy an airplane if you can't fly it?

Giving One to Each

When things are shared equally, everyone or everything involved gets an equal share — just one of the shares — not twice as many as others get. When a child is distributing her birthday treats to classmates, it's: "One for you, and one for you . . ." In the game Mancala, the stones in a cup are distributed one to each of the next cups until they're gone. Any other way is cheating! When the prize money is parceled out to the winning team members, the shares are distributed to each member. In algebra, distributing is much the same process — each gets a share.

Distributing items is an act of spreading them out equally. Algebraic distribution means to multiply each of the terms within the parenthesis by another term that is outside the parenthesis.

To distribute a term over several other terms, multiply each of the other terms by the first. Distribution is multiplying each individual term in a grouped series of terms by a term outside of the grouping.

$$a(b+c) = a \cdot b + a \cdot c$$
$$a(b-c) = a \cdot b - a \cdot c$$

A *term* is made up of variable(s) and/or number(s) joined by multiplication and/or division. Terms are separated from one another by addition and/or subtraction.

1. **Multiply each term by the number(s) and/or variable(s) outside of the parenthesis.**

   Distribute the number two over the terms in the parenthesis.

   $$2(4x + 3y - 6) =$$
   $$2(4x) + 2(3y) - 2(6)$$

2. **Perform the multiplication operation in each term.**

   $$8x + 6y - 12.$$

Now that you have the idea, try extracting a distribution operation out of the following scenario: At a particular car dealership, five salespersons, A, B, C, D, and E, sold 2, 8, 6, 5, and 9 cars last month, respectively. This is $2 + 8 + 6 + 5 + 9 = 30$ cars. The owner of the dealership wants to double the sales this month. He wants his sales force to sell a total of 60 cars. If each salesperson doubles what he sold last month, look at what happens.

1. **Multiply each term by the number(s) and/or variable(s) outside of the parenthesis.**

   $$2(2 + 8 + 6 + 5 + 9) = 2(2) + 2(8) + 2(6) + 2(5) + 2(9)$$

2. **Perform the multiplication operation and add each term.**

   $$4 + 16 + 12 + 10 + 18 = 60 \text{ cars}$$

The answer is the same, of course, whether you distribute first or add up what's in the parenthesis first. You have to make that judgment call. Distributing first to get the answer is the better choice when the multiplication of each term gives you nicer numbers. Fractions or decimals in the parenthesis are sometimes changed into nice whole numbers when the distribution is done first. The other choice, adding up what's in the parenthesis first, is preferred when the distributing gives you too many big multiplication problems. Sometimes it's easy to tell which case you have. At other times, you just have to guess and try it.

Distributing first

First, take a moment to look at the distribution problem before you. Sometimes you can tell just by looking whether it's easier to distribute the term outside the parenthesis before or after summing up the terms within the parenthesis. Taking a moment to see what may be the best approach may save time in the long run.

The following example shows you what it means to make a judgment call to distribute first.

✔ Look at what's involved if you distribute 60 over $\frac{1}{2} + \frac{3}{5} - \frac{3}{4} + \frac{13}{15}$ by adding the fractions first.

Find a common denominator and then add and subtract the fractions.

$$60\left(\frac{30}{60} + \frac{36}{60} - \frac{45}{60} + \frac{52}{60}\right) =$$

$$60\left(\frac{73}{60}\right) = 73$$

✔ Now look at the better choice, where the distribution is done first.

$$60\left(\frac{1}{2} + \frac{3}{5} - \frac{3}{4} + \frac{13}{15}\right)$$

Multiplying by 60 gets rid of all the fractions, so you don't have to find a common denominator.

$$60\left(\frac{1}{2}\right) + 60\left(\frac{3}{5}\right) - 60\left(\frac{3}{4}\right) + 60\left(\frac{13}{15}\right) =$$
$$30 + 36 - 45 + 52 = 73$$

Do you see the advantage in this case of doing the distribution first? In the following case, however, combining what's in the parenthesis first is easier.

Adding first

Before working through a distribution problem, look at the size of the numbers. If the numbers are large, then distributing one large term over other large terms within the parenthesis can only make each term larger and less manageable. In the case of large numbers, it may be easier to work through any simple addition and subtraction within the parenthesis before distributing the term outside the parenthesis over those within.

✔ Distribute 43 first (ugh).

$$43(160 - 159 + 433 - 432) =$$

$$43(160) - 43(159) + 43(433) - 43(432) =$$

$$6880 - 6837 + 18{,}619 - 18{,}576 =$$

86

✔ Now look at the better choice, where you combine first.

$$43(160 - 159 + 433 - 432) =$$

$$43(1+1) = 43(2) = 86$$

These last two examples are a bit exaggerated. The best route to take isn't always obvious. But if you keep your eyes open for the choices available, you can save yourself some time and work.

Distributing Signs

A positive (+) or a negative (−) sign is simple to distribute, but distributing a negative sign can cause *errors*. More contradictions!

Distributing positives

Distributing a positive sign makes no difference in the signs of the terms.

✔ $+(4x + 2y - 3z + 7)$ is the same as multiplying through by +1.

$$+1(4x + 2y - 3z + 7) = +1(4x) + 1(2y) + 1(-3z) + 1(7) =$$

$$4x + 2y - 3z + 7$$

Even when a positive number other than the number one is distributed, it doesn't affect the signs.

✔ $+3(4x + 2y - 3z + 7) = +3(4x) + 3(2y) + 3(-3z) + 3(7) =$

$$12x + 6y - 9z + 21$$

It should come as no big surprise that the signs of the expression stayed the same.

Distributing negatives

When distributing a negative sign, each term has a change of sign from negative to positive or from positive to negative.

Distribute a negative sign through a bunch of terms.

✔ $-(4x + 2y - 3z + 7)$ is the same as multiplying through by -1.

$-1(4x + 2y - 3z + 7) = -1(4x) - 1(2y) - 1(-3z) - 1(7) =$

$-4x - 2y + 3z - 7$

Change each term to the opposite sign.

WARNING!

One mistake to avoid when you're distributing a negative sign is not distributing over *all* the terms. This is especially the case when the process is *hidden*. By hidden, I mean that a negative sign may not be in front of the whole expression, where it sticks out. It can be between terms, showing a subtraction and not being recognized for what it is. Don't let the negative signs ambush you.

1. **Distribute the term.**

   $4x(x - 2) - (5x + 3)$

   Distribute the $4x$ over the x and the 2 by multiplying both terms by $4x$

   $4x(x - 2) = 4x(x) - 4x(2)$

2. **Distribute any negative sign.**

   Distribute the negative sign over the $5x$ and the 3 by changing the sign of each term. Be careful; a mistake can easily be made when the negative is distributed only over the $5x$. $-(5x + 3) = -(+5x) - (+3)$

3. **Multiply and combine the like terms.**

   $4x(x) - 4x(2) - (+5x) - (+3) = 4x^2 - 8x - 5x - 3 =$

   $4x^2 - 13x - 3$

FUN FACT

Palindromes

The word *palindrome* comes from the Greek word *palindromos,* which means *running back again.* A palindrome is any word, sentence, or even a complete poem that reads the same backwards as it does forwards. For example, Leigh Mercer wrote "A man, a plan, a canal — Panama" to honor the man responsible for building the Panama Canal. Or, how about, "Niagara, O roar again!" There are words that are palindromes: *rotator, Malayalam* (East Indian language), and *redivider.*

Number palindromes have been of great interest to mathematicians over the years. The perfect squares are palindromes: 121 and 14,641 for example. A palindromic day might be October 9th, 1901 (1091901). Apparently, reversing the digits of almost any number and adding the reversal to the original number can create a palindrome. If not, however, just repeat the steps until you get a palindrome. For example, take 146, reverse the digits to get 641. Add them together: 146 + 641 = 787.

Reversing the roles in distributing

Distributing multiplication over an expression that has several terms added or subtracted is an extension of simply multiplying. What does this do in terms of the *commutative* law?

Multiplication is commutative, which means that the order in which you multiply the terms doesn't matter: $a \times b = b \times a$.

What happens to distributing if you reverse the order? After all, adding and subtracting is involved, too.

1. **Distribute the term in front of the parenthesis over each term.**

   Consider $3(x^2 + y - 7 - z)$ and compare it to $(x^2 + y - 7 - z)3$. In the first expression, the three is in front.

   $$3(x^2 + y - 7 - z) = 3(x^2) + 3(y) + 3(-7) + 3(-z)$$

2. **Multiply each term.**

   $$3x^2 + 3y - 21 - 3z$$

3. **Distribute each term over the term after the parenthesis.**

   In the second expression, the three is in back.

   $$(x^2 + y - 7 - z)3 = x^2(3) + y(3) - 7(3) - z(3)$$

   The factors can be reversed in each of the terms because multiplication is commutative.

4. **Multiply and rewrite.**

   $$3x^2 + 3y - 21 - 3z$$

   The results are exactly the same. Hurrah! A task made easier.

Mixing It Up with Numbers and Variables

Distributing variables over the terms in an algebraic expression involves multiplication rules and the rules for exponents. When different variables are multiplied together, they can be written side by side without using any multiplication symbols. If the same variable is multiplied as part of the distribution, then the exponents are added together.

The exponent rule says that when multiplying exponents with the same base, add the exponents:

$$a^2 \cdot a^3 = a^5.$$

The exponent rule for multiplying terms with the same base is used in the following problem.

1. **Distribute the term outside the parenthesis over the terms within.**

   Multiply a through the expression $a^4 + 2a^2 + 3$

   $$a(a^4 + 2a^2 + 3) = a \cdot a^4 + a \cdot 2a^2 + a \cdot 3$$

2. **Complete the multiplication.**

   $$a^5 + 2a^3 + 3a$$

And, again:

1. **Distribute the term outside the parenthesis over the terms within.**

   Multiply a^3 through the expression $a^4 + 2a^2 + 3$

   $$a^3(a^4 + 2a^2 + 3) = a^3 \cdot a^4 + a^3 \cdot 2a^2 + a^3 \cdot 3$$

2. **Add the exponents.**

   $$a^{3+4} + 2a^{2+3} + a^3 \cdot 3$$

3. **Complete the distribution.**

   $$a^7 + 2a^5 + 3a^3$$

Adding exponents doesn't change just because there are negatives or fractions. For example: $4 + (-2) = 2$; $4 + (-4) = 0$; $4 + \frac{1}{3} = 4\frac{1}{3} = \frac{13}{3}$.

1. **Distribute the term outside the parenthesis over those within.**

   Distribute the z^4 over each term.

   $$z^4(2z^2 - 3z^{-2} + z^{-4} + 5z^{1/3}) = z^4 \cdot 2z^2 - z^4 \cdot 3z^{-2} + z^4 z^{-4} + z^4 \cdot 5z^{1/3}$$

2. **Add the exponents.**

   $$2z^{4+2} - 3z^{4-2} + z^{4-4} + 5z^{4+1/3}$$

3. **Complete the distribution.**

   $$2z^6 - 3z^2 + z^0 + 5z^{13/3} = 2z^6 - 3z^2 + 1 + 5z^{13/3}$$

The exponent zero means the value of the expression is one.

$$x^0 = 1$$

You combine exponents with different signs by using the rules for adding and subtracting signed numbers. Fractional exponents are combined after finding common denominators. Exponents that are improper fractions are left in that form.

Try going through many of the situations that could arise when distributing, such as distributing both a number and a variable, distributing various powers of more than one variable, distributing negatives, rewriting negative exponents as fractional terms, distributing fractional powers, and distributing radicals. This touches on just about anything you'd be apt to come across.

✔ Combine the variables by using the rules for exponents.

Multiply each term by $5x$.

$$5x(2x^2 + 3x - 4) = 5x \cdot 2x^2 + 5x \cdot 3x - 5x \cdot 4$$

Multiply the numbers and the variables in each term.

$$10x^3 + 15x^2 - 20x$$

✔ Combine the variables with the same base using the rules for exponents. The signs of the results follow the rules for multiplying signed numbers.

$$-6y(5xy - 4x - 3y + 2)$$

Multiply each term by $-6y$.

$$-6y(5xy) - 6y(-4x) - 6y(-3y) - 6y(2)$$

Do the multiplication in each term.

$$-30xy^2 + 24xy + 18y^2 - 12y$$

✔ Notice that the last term in the next answer is the opposite of the term outside the parenthesis. It is multiplied by the -1, which is the last term within the parenthesis.

$$5x^2 y^3 (16x^2 - 2x + 3xy + 4y^3 - 11y^5 + z - 1) =$$

Multiply each term by $5x^2 y^3$.

$$5x^2 y^3 \cdot 16x^2 - 5x^2 y^3 \cdot 2x + 5x^2 y^3 \cdot 3xy + 5x^2 y^3 \cdot 4y^3 - 5x^2 y^3 \cdot 11y^5$$
$$+5x^2 y^3 z - 5x^2 y^3$$

Complete the multiplication in each term. Add exponents where needed.

$$80x^4 y^3 - 10x^3 y^3 + 15x^3 y^4 + 20x^2 y^6 - 55x^2 y^8 + 5x^2 y^3 z - 5x^2 y^3$$

There are no like terms to be combined.

Distributing negative signs

Distribute the negative sign through, changing the signs to their opposites. Only the variables that are alike have exponential changes because the bases are the same.

✔ Look at the following problem, which is cluttered with negative signs.

$$-4xyzw(4 - x - y - z - w)$$

Multiply each term by $-4xyzw$.

$$-4xyzw(4) - 4xyzw(-x) - 4xyzw(-y) - 4xyzw(-z) - 4xyzw(-w)$$

Complete the multiplication in each term.

$$-16xyzw + 4x^2 yzw + 4xy^2 zw + 4xyz^2 w + 4xyzw^2$$

✔ In this next example, the $-5x$ distributes over the first two terms. The $4x^2$ distributes over the second two terms.

$$-5x(3x + 2) + 4x^2(5 - 3x)$$

Distribute the two different factors.

$$-5x \cdot 3x - 5x \cdot 2 + 4x^2 \cdot 5 + 4x^2(-3x)$$

Multiply the individual terms.

$$-15x^2 - 10x + 20x^2 - 12x^3$$

The two terms with x^2 combine into one term.

$$-10x + 5x^2 - 12x^3$$

Negative exponents yield fractional answers

As the heading suggests, a base that has a negative exponent can be changed to a fraction. The base and the exponent become the denominator, but the exponent loses its negative sign in the process. Then cap it all off with a one in the numerator.

The formula for changing negative exponents to fractions is $a^{-n} = \frac{1}{a^n}$. See Chapter 4 for more details on negative exponents.

The following example shows you how a negative exponent can lead to a fraction.

✔ Distribute the $5a^{-3}b^{-2}$ over each term in the parenthesis.

$$5a^{-3}b^{-2}(2ab^3 - 3a^2b^2 + 4a^4b - ab) =$$

$$5a^{-3}b^{-2}(2ab^3) - (5a^{-3}b^{-2})(3a^2b^2) + (5a^{-3}b^{-2})(4a^4b) - (5a^{-3}b^{-2})(ab)$$

Add the exponents.

$$10a^{-3+1}b^{-2+3} - 15a^{-3+2}b^{-2+2} + 20a^{-3+4}b^{-2+1} - 5a^{-3+1}b^{-2+1}$$

The factor of b with the 0 exponent becomes 1.

$$10a^{-2}b^1 - 15a^{-1}b^0 + 20a^1b^{-1} - 5a^{-2}b^{-1}$$

This next step shows the final result without negative exponents — using the formula for changing negative exponents to fractions, which is stated earlier in this section.

$$\frac{10b}{a^2} - \frac{15}{a} + \frac{20a}{b} - \frac{5}{a^2b}$$

Working with fractional powers

Exponents that are fractions work the same way as exponents that are integers. They're added together. The only hitch is that the fractions have to have the same denominator to be added.

✔ Try the following distribution problem with exponents that are fractions to give yourself an idea of how to work through it.

$$x^{1/4}y^{2/3}\left(x^{1/2} + x^{3/4}y^{1/3} - y^{-1/3}\right)$$

Distribute the $x^{1/4}y^{2/3}$.

$$x^{1/4}y^{2/3} \cdot x^{1/2} + x^{1/4}y^{2/3} \cdot x^{3/4}y^{1/3} - x^{1/4}y^{2/3} \cdot y^{-1/3}$$

Rearrange the variables and add exponents.

$$x^{1/4}x^{1/2}y^{2/3} + x^{1/4}x^{3/4}y^{2/3}y^{1/3} - x^{1/4}y^{2/3}y^{-1/3}$$

$$x^{1/4+1/2}y^{2/3} + x^{1/4+3/4}y^{2/3+1/2} - x^{1/4}y^{2/3-1/3}$$

Add the fractions.

$$x^{3/4}y^{2/3} + x^1y^1 - x^{1/4}y^{1/3}$$

Radicals can be changed to expressions with fractions as exponents. This is handy when you want to combine terms with the same bases and you have some of the bases under radicals.

✔ $\sqrt{x} = x^{1/2}$

✔ $\sqrt{xy} = \sqrt{x}\sqrt{y} = x^{1/2}y^{1/2}$

✔ $\sqrt{x^3} = (x^3)^{1/2} = x^{3/2}$

Distribution is easier in this case if you first change everything to fractional exponents. (See more on exponential operations within radicals in Chapter 4.)

The root of the radical becomes the denominator of the fractional exponent. For example: $\sqrt[n]{a} = a^{1/n}$ and $\sqrt[n]{a^m} = a^{m/n}$.

The exponent rule for raising a product in a parenthesis to a power is to multiply each power in the parenthesis by the outside power. For example: $(x^4 y^3)^2 = x^8 y^6$.

1. **Change the radical notation to fractional exponents.**

$$\sqrt{xy^3}\left(\sqrt{x^5 y} - \sqrt{xy^7}\right) = (xy^3)^{1/2}\left[(x^5 y)^{1/2} - (xy^7)^{1/2}\right]$$

2. **Raise the powers.**

$$x^{1/2} y^{3/2}\left[x^{5/2} y^{1/2} - x^{1/2} y^{7/2}\right]$$

3. **Distribute the outside term over each term within the parenthesis.**

$$x^{1/2} y^{3/2}(x^{5/2} y^{1/2}) - x^{1/2} y^{3/2}(x^{1/2} y^{7/2})$$

4. **Add the exponents of the variables.**

$$x^{6/2} y^{4/2} - x^{2/2} y^{10/2}$$

5. **Simplify the fractions.**

$$x^3 y^2 - x^1 y^5$$

To *simplify* means to combine all that can be combined to put an expression in its most easily understood form.

In this example, the only thing accomplished is that the usual conventions of having the number first, followed by variables in alphabetical order are observed.

✔ As you see, nothing combines in the following example.

$$(4 - 5x^2 y + mnp)$$

Distribute $a^3 b^4 c^6$.

$$a^3 b^4 c^6(4) - a^3 b^4 c^6(5x^2 y) + a^3 b^4 c^6(mnp)$$

Multiply each term.

$$4a^3 b^4 c^6 - 5a^3 b^4 c^6 x^2 y + a^3 b^4 c^6 mnp$$

Distributing More than One Term

The preceding sections in this chapter describe how to distribute one term over several others. This section shows you how to distribute a *binomial,* a polynomial with two terms. You'll also discover how to distribute polynomials with three or more terms.

The highest power in a polynomial

What is the highest power in a polynomial? The highest power of $x^3 + 1$, for example, is 3, so it's a *cubic binomial.* The word *cubic* tells you that the highest power is 3, and the word *binomial* tells you that the polynomial has a total of 2 terms, the x^3 and the 1. A *quintic* polynomial has a highest power of 5.

The following italicized words are listed alongside their corresponding powers. The phrases in parenthesis that follow each word in the list can help you to remember and associate the word with the power that it refers to.

✔ First power: linear or *monic* (as in *mono*gamy, for one wife)

✔ Second power: *quadratic* (*quad,* for a four-sided square)

✔ Third power: *cubic* (a sugar cube with three dimensions.)

✔ Fourth power: *quartic* (four cups in a *quart*)

✔ Fifth power: *quintic* (*quint*uplets — five of them!)

✔ Sixth power: *sextic* (or, if this is too risqué, then use *hexic*)

✔ Seventh power: *septic* (as opposed to sewers — or, if you prefer, *heptic*)

✔ Eighth power: *octic* (eight-legged octopus)

✔ Ninth power: *nonic* (*non*-ic: not icky?)

✔ Tenth power: *decic* (as in *deci*mal and *deci*bel)

✔ Hundredth power: *hectic* (That's for sure!)

The word *polynomial* comes from *poly* meaning *many,* and *nomen* meaning *name* or *designation.* A polynomial is an algebraic expression with one or more terms in it. For example, a polynomial with one term is a *monomial;* a polynomial with two terms is a *binomial.* If there are three terms, it's a *trinomial.*

Distributing binomials

Distributing two terms, or a *binomial,* over several terms amounts to just applying the distribution process twice. The following steps tell you how:

1. **Break the first binomial into its two terms.**

   In this case, $(x^2 + 1)(y - 2)$, break the first binomial into its two terms, x^2 and 1.

2. **Distribute each term of the first binomial over the other terms.**

   Distribute the first term, which is x^2, of first binomial $(x^2 + 1)$, over the second binomial, and distribute the second term, which is 1, of the first binomial over the second binomial.

   $x^2(y - 2) + 1(y - 2)$

3. Do the two distributions.

$$x^2(y-2)+1(y-2)=x^2y-2x^2+y-2$$

4. Simplify and combine any like terms.

In this case, nothing can be combined.

Now that you have the idea, try walking through a polynomial distribution that has variables in all the terms.

1. Break the first binomial into its two terms.

$$(a+b)(a^2-ab+b^2)$$

Break the first binomial into its two terms to distribute it over the other terms.

2. Distribute each term of the first binomial over the other terms.

Distribute a and b over the other terms.

$$a(a^2-ab+b^2)+b(a^2-ab+b^2)$$

3. Do the two distributions.

$$a^3-a^2b+ab^2+a^2b-ab^2+b^3$$

4. Simplify and combine any like terms.

Some terms can be combined. Note that the second and fourth terms are opposites and that the third and fifth terms are opposites.

$$a^3+b^3$$

This simplifies nicely.

As you work through the example that follows, notice that the negative sign distributed over the last three terms and that the middle two terms combine.

1. Break the binomial into its two terms x^2 and $-y^2$.

$$(x^2-y^2)(x^2+2xy+y^2)$$

2. Distribute each term of the binomial over the other terms.

$$x^2(x^2+2xy+y^2)-y^2(x^2+2xy+y^2)$$

3. Do the two distributions.

$$x^2(x^2)+x^2(2xy)+x^2(y^2)-y^2(x^2)-y^2(2xy)-y^2(y^2)$$

4. Simplify and combine any like terms.

Multiply and add exponents.

$$x^4+2x^3y+x^2y^2-x^2y^2-2xy^3-y^4$$

Combine terms.

$$x^4+2x^3y-2xy^3-y^4$$

Distributing trinomials

A trinomial, a polynomial with three terms, can be distributed over another expression. Each term in the first factor is distributed separately over the second factor, and then the entire expression is simplified, combining anything that can be combined.

The following problem introduces you to working through the distribution of trinomials.

1. **Break the trinomial into its three terms x, y, and 2.**

   $$(x + y + 2)(x^2 - 2xy + y + 1)$$

2. **Distribute each term of the trinomial over the other terms.**

   $$x(x^2 - 2xy + y + 1) + y(x^2 - 2xy + y + 1) + 2(x^2 - 2xy + y + 1)$$

3. **Do the three distributions.**

   $$x^3 - 2x^2 y + xy + x + x^2 y - 2xy^2 + y^2 + y + 2x^2 - 4xy + 2y + 2$$

4. **Simplify and combine any like terms.**

   $$x^3 - x^2 y + 2x^2 + x - 2xy^2 + y^2 - 3xy + 3y + 2$$

Trinomial times a polynomial

This is where you can establish a rule that can cover just about any product of any number of terms. You can use this general method for four, five, or even more terms.

When distributing a polynomial (many terms) over any number of other terms, distribute each term in the first factor over all of the terms in the second factor. When the distribution is done, combine anything that goes together to simplify.

The following example is made up of nothing but variables, none of which are the same.

1. **Separate the terms in the first factor from one another. Multiply each term in the first factor times the second factor.**

   $$(a + b + c + d + \cdots)(z + y + w + x + \cdots) =$$

   $$a(z + y + x + w + \cdots) + b(z + y + x + w + \cdots) + c(z + y + x + w + \cdots) + \cdots$$

2. **Distribute and do the multiplication.**

   $$az + ay + ax + aw + \cdots + bz + by + bx + bw + \cdots + cz + cy + cx + cw + \cdots$$

3. Combine like terms.

In this case, none of the terms are alike, but you should check.

The next example shows you how to multiply two trinomials.

1. **Separate the terms in the first factor from one another. Multiply each term in the first factor times the second factor.**

$$(x^2 + x + 2)(3x^2 - x + 1)$$

$$x^2(3x^2 - x + 1) + x(3x^2 - x + 1) + 2(3x^2 - x + 1)$$

2. **Distribute and do the multiplication.**

$$3x^4 - x^3 + x^2 + 3x^3 - x^2 + x + 6x^2 - 2x + 2$$

3. **Combine like terms.**

$$3x^4 + 2x^3 + 6x^2 - x + 2$$

Like the expression that results in the difference of two cubes, the problem that follows reuses the same two variables. Further, the first factor is a binomial, and the second factor is a trinomial. But this problem is a little different.

1. **Separate the terms in the first factor from one another. Multiply each term in the first factor times the second factor.**

$$(a^2 - b^2)(a^2 - ab - b^2)$$

$$a^2(a^2 - ab - b^2) - b^2(a^2 - ab - b^2)$$

2. **Distribute and do the multiplication.**

$$a^4 - a^3b - a^2b^2 - a^2b^2 + ab^3 + b^4$$

Notice the care taken when the second, negative factor of the binomial was multiplied through.

3. **Combine like terms.**

$$a^4 - a^3b - 2a^2b^2 + ab^3 + b^4$$

Making Special Distributions

Several distribution shortcuts can make life easier. Distributing binomials over other terms is not difficult, but you can save time if you recognize problems where you can apply a shortcut. If you don't notice that a special shortcut could have been used, don't worry about your oversight. But you may end up kicking yourself afterwards for not taking advantage of the easier process.

Recognizing the perfectly squared binomial

When the same binomial is multiplied by itself — when each of the first two terms is distributed over the second and same terms — then the resulting trinomial contains the *squares* of the two terms and twice their product:

$$(a+b)^2 = (a+b)(a+b) = a^2 + 2ab + b^2$$

✔ The result of the following operation is the sum of the squares of x and 3 along with twice their product.

$$(x+3)(x+3)$$

The square of x is x^2

The square of 3 is 9

Twice the product of x and 3 is $2(3x) = 6x$

$$(x+3)(x+3) = x^2 + 6x + 9.$$

Notice that the usual order of the terms was used: decreasing powers of x.

✔ Try the following binomial distribution with negative signs. Don't forget to square both the 4 and the y.

$$(4y-5)(4y-5)$$

The square of $4y$ is $16y^2$.

Note that the next square is positive.

The square of -5 is $+25$.

Twice the product of $4y$ and -5 is $2(4y)(-5) = -40y$

So $(4y-5)(4y-5) = 16y^2 - 40y + 25$

✔ In the following example, the terms are all variables.

$$(a^3 + b^2)(a^3 + b^2)$$

The square of a^3 is $(a^3)^2 = a^6$

The square of b^2 is $(b^2)^2 = b^4$

Twice the product of a^3 and b^2 is $2a^3b^2$

So $(a^3 + b^2)(a^3 + b^2) = a^6 + 2a^3b^2 + b^4$

✔ Parentheses group the last two terms together in this trinomial distribution.

$$[x + (a+b)][x + (a+b)]$$

The square of x is x^2

The square of $(a+b)$ is $(a+b)^2 = a^2 + 2ab + b^2$

Twice the product of x and $(a + b)$ is $2x(a + b)$

$$[x + (a + b)][x + (a + b)] = x^2 + 2x(a + b) + a^2 + 2ab + b^2$$

If you want to further multiply this out, you can distribute that second term.

$$x^2 + 2xa + 2xb + a^2 + 2ab + b^2$$

Spotting the sum and difference of the same two terms

There's just one little — which can be *big* — difference between these multiplications and the ones in the previous section. The difference is that there's a sign change between the first and second binomials. If the sign between the two terms in the first binomial is positive, then in the second it's negative. The same two terms are always used — it's just that the sign between them changes.

The sum of any two terms multiplied by the difference of the same two terms is easy to spot and even easier to work out.

The sum of any two terms multiplied by their difference equals the difference of the squares of the same two terms. For any real numbers a and b:

$$(a + b)(a - b) = a^2 - b^2$$

Notice that the middle term just disappears because a term and its opposite are always in the middle. You can see that here:

$$(a + b)(a - b) = a(a - b) + b(a - b)$$

Distribute the a and b over the second factor.

$$a^2 - ab + ab - b^2 = a^2 - b^2$$

✔ The rule always works, so you can use the shortcut to do these special distributions.

$$(x - 4)(x + 4)$$

The first term squared is x^2

The second term will always be negative and a perfect square like the first term.

$$(-4)(+4) = -16$$

So $(x - 4)(x + 4) = x^2 - 16$

✔ Try the same easy process — multiplying the sum of two terms with their difference — again with a slightly more complicated variable term.

$$(ab - 5)(ab + 5)$$

The square of $ab = (ab)^2 = a^2 b^2$

The opposite of the square of $5 = -25$

So $(ab - 5)(ab + 5) = a^2 b^2 - 25$

✔ The following example offers you a chance to work through the sum and difference of various groupings.

$$[5 + (a - b)][5 - (a - b)]$$

The square of $5 = 25$

The opposite of the square of $(a - b) = -(a - b)^2$

Square the binomial and distribute the negative sign.

$$-(a^2 - 2ab + b^2) = -a^2 + 2ab - b^2$$

So $[5 + (a - b)][5 - (a - b)] = 25 - a^2 + 2ab - b^2$

Working out the difference between two cubes

So, what are *cubes?* Although some cubes are made of sugar and spice and everything nice, the cubes used in algebra are slightly different. Some of them are three-dimensional objects, but the cubes in this section are values that are multiplied times themselves — *three* times. Remember, a value multiplied by itself is a perfect square; a value multiplied by itself three times is a perfect cube. The variable x cubed is written: x^3. For example, three cubed (3^3) is 27 because $3 \times 3 \times 3 = 27$.

An expression that results in the difference between two cubes is usually pretty hard to spot. You may not notice it until you get to the final answer and then say, "Oh, yeah. That's right!" However, being able to recognize what results in the difference of two cubes is even more important in Chapter 15, which addresses cubic equations (equations that contain a term with an exponent of three and no higher).

The difference of two cubes is equal to the difference of their cube roots times a trinomial, which contains the squares of the cube roots and the opposite of the product of the cube roots. For any real numbers a and b,

$$(a - b)(a^2 + ab + b^2) = a^3 - b^3$$

To recognize what distribution results in the difference of two cubes, look to see if the distribution has a binomial, $(a - b)$, which is the difference between two terms, times a trinomial, $(a^2 + ab + b^2)$, which has the squares of the two terms and the opposite their product.

A number's *opposite* is that same number with a different sign in front. If the number is a negative number, then its opposite would be positive and vice versa.

✔ Go ahead and distribute to see why this works.

$$(a - b)(a^2 + ab + b^2)$$

Distribute the a and the $-b$ over the trinomial.

$$a(a^2 + ab + b^2) - b(a^2 + ab + b^2)$$

Distribute the two values separately and multiply each term.

$$a^3 + a^2 b + ab^2 - a^2 b - ab^2 - b^3$$

Notice that the four terms in the middle are all pairs of opposites that add up to zero.

Combine like terms.

$$a^3 - b^3$$

This pattern always results in the difference of two cubes.

✔ If you recognize the pattern in the following example, which has terms comprising known numbers and multiple variables, then consider yourself an algebraic gold medalist!

$$(2 - ab)(4 + 2ab + a^2 b^2)$$

The square of ab is $(ab)^2 = a^2 b^2$, and the cube of ab is $(ab)^3 = a^3 b^3$.

So $(2 - ab)(4 + 2ab + a^2 b^2) = 8 - a^3 b^3$.

Finding the sum of two cubes

This should look familiar. It's just like working out the result of the difference of two cubes, except that two signs change. The binomial has a plus sign and the middle term in the trinomial is minus.

$$(a + b)(a^2 - ab + b^2) = a^3 + b^3$$

When working with the two factors, the binomial and the trinomial, that give you the sum or difference of two cubes, the only difference in the factors is the two signs. The sign in the binomial is always the opposite of the sign in the middle of the trinomial. Look at what I mean in the following equations. The first result is the sum of two cubes, and the second result is the difference of two cubes.

$$(a + b)(a^2 - ab + b^2) = a^3 + b^3$$
$$(a - b)(a^2 + ab + b^2) = a^3 - b^3$$

Work through the following examples to give yourself a better understanding:

✔ The sign in the binomial is +, so the answer has a +. The cube of 4 is 64.

$$(x+4)(x^2-4x+16)=x^3+64$$

✔ The opposite of the product of 6 and $5yz$ is $-30yz$. The cube of $5yz = (5yz)^3 = 5^3\,y^3\,z^3$.

$$(6+5yz)(36-30yz+25y^2\,z^2)=216+125y^3\,z^3$$

Just as it's nice to have a list of perfect squares because they're used so much in algebra, it's equally nice to have a list of perfect cubes. You can always use your calculator to find the cube of a number, but it saves time if you already know it. Recognizing that a number is a perfect cube can come in handy later.

Try committing the list in Table 8-1 to memory.

Table 8-1	First Ten Perfect Cubes
$1^3 = 1$	$6^3 = 216$
$2^3 = 8$	$7^3 = 343$
$3^3 = 27$	$8^3 = 512$
$4^3 = 64$	$9^3 = 729$
$5^3 = 125$	$10^3 = 1{,}000$

Chapter 9

Factoring in the First Degree

You may believe in the bigger-the-better philosophy, which can apply to salaries, cookies, or houses, but it doesn't really work for algebra. For the most part, the opposite is true: Smaller numbers are easier and more comfortable to deal with than larger numbers. In this chapter, you can discover how to get to those smaller-is-better terms. First-degree terms have a variable with an exponent of one. The factoring patterns you see here will carry over somewhat in higher degrees.

Factoring

Factoring is another way of saying: "Rewrite this so everything is all multiplied together." You usually start out with two or more terms and have to determine how to rewrite them so they're all multiplied together in some way or another. And, oh yes, the two expressions have to be equal!

Factoring out numbers

Before I start giving you rules and instructions on how to factor in algebra, you may want to see what the results of factoring look like. Take a look at the following examples, first:

Factoring is the opposite of distributing; it's "undistributing" (see Chapter 8 for more on distribution). In distribution, you multiplied a series of terms by a common factor. Now, by factoring, you seek to find what a series of terms have in common and then take it away, dividing the common factor out from each term. Think of each term as a numerator and then find the same denominator for each. By factoring *out*, the common factor is put *outside* a parenthesis or bracket and all the results of the divisions are left inside.

1. **Determine a common factor.**

   In the term $16a - 8b + 40c^2$, 2 is a common factor.

2. **Divide or "undistribute" each term by the common factor and write the results of the division in parenthesis with the factor out in front.**

   In this example, this looks like $16a - 8b + 40c^2 = 2(8a - 4b + 20c^2)$

3. **Determine whether you can factor out any other terms.**

   The terms left in the parenthesis are still too big. They all still have something in common: 4. Factoring out 4, you get

   $$2(4[2a - b + 5c^2])$$

4. **Simplify your answer.**

   If you factor out a 4 after factoring out the 2, then the product of 4 and 2, which is 8, is the total amount you factored out:

   $$8(2a - b + 5c^2)$$

It's nice when you recognize going in that you can factor out a larger number, such as 8, but if it takes a couple steps to get to it, that's fine.

Reviewing the definitions of the following words and keeping them in mind can help.

- ✔ **Term:** Group of number(s) and/or variable(s) connected to one another by multiplication or division and separated from other terms by addition or subtraction.

- ✔ **Factor:** Any of the values involved in a multiplication problem that when multiplied together produces a result.

- ✔ **Coefficient:** Number that multiplies a variable and tells how many of the variable.

- ✔ **Constant:** Number or variable that never changes in value.

For example, in the expression $5xy + 4z - 6$, three terms are separated by the plus and minus signs. In the first term, $5xy$, three factors are all multiplied together. The 5 is usually referred to as the coefficient. The second term has two factors, 4 and z, and the third term has only one — a constant.

An expression can be written as the product of the largest number that divides all the terms evenly times the results of the divisions.

$$ab + ac + ad = a(b + c + d)$$

The following examples put this to practice:

✔ Stephen has 6 cats; Brad has 18 hamsters; Carlos has 16 parakeets; Donald has 4 dogs. These pet owners want to take their pets to various nursing homes to visit the residents, but they want to divide the animals into similar groups. How can they do this?

The sum of numbers representing the animals is 6 + 18 + 16 + 4, each of which can be divided evenly by 2. The 6 and 18 can be divided by 6, but the 16 and 4 cannot be divided by 6. The 16 and 4 can be divided by 4, but the six and 18 cannot be divided by 4.

Two is the biggest number that divides each evenly.

So these gentlemen can take 2 groups of animals to the nursing homes: 2(3 cats + 9 hamsters + 8 parakeets + 2 dogs). What a nice thing for them to do!

✔ Each of the terms in this example has a coefficient that three divides evenly. The GCF (greatest common factor) of the numbers is three, so a factor larger than three that can divide *all* the terms evenly is unavailable.

Factor $9x + 15y - 12z + 30$:

$$9x + 15y - 12z + 30 = 3(3x + 5y - 4z + 10)$$

The terms in the parenthesis are the results of dividing each term by 3. Those terms don't have anything in common.

✔ Each term in the following example is divisible by six.

Factor $18a^2 - 24b - 36c + 42$

$$18a^2 - 24b - 36c + 42 = 6(3a^2 - 4b - 6c + 7)$$

Relatively prime means that two terms have no prime factors in common. If the only factor that two numbers share in common is 1, then they're considered *relatively prime.*

For example, one is the only number that divides into both 18 and 15. Although neither 18 nor 25 is a prime number, they are *relatively prime.*

✔ The *proper* way to factor the following expression would be to write the prime factorization of each of the numbers and look for the GCF (greatest common factor). What's really more practical and quicker in the end is to look for the biggest factor that *you can easily recognize.* Factor it out and then see if the numbers in the parentheses need to be factored again. Repeat the division until the terms in the parentheses are relatively prime.

$$450x + 540y - 486z + 216$$

Divide each term by two.

$$450x + 540y - 486z + 216 = 2(225x + 270y - 243z + 108)$$

The numbers in the parentheses are a mixture of odd and even, so you can't divide by two again. The numbers in the parentheses *are* all divisible by three, but there's an even better choice.

You may have noticed that the digits in the numbers in all the terms add up to nine. That's the rule for divisibility by nine, so nine can divide each term evenly. Thus,

$$2(225x + 270y - 243z + 108) = 2[9(25x + 30y - 27z + 12)]$$

Now multiply the two and nine together to get

$$450x + 540y - 486z + 216 =$$

$$18(25x + 30y - 27z + 12)$$

You could have divided 18 into each term in the first place, but not many people know the multiplication table of 18. It's a stretch even for me.

Factoring out variables

Variables represent values; variables with exponents represent the powers of those same values. For that reason, variables as well as numbers can be factored out of the terms in an expression, and in this section you can find out how.

When factoring out powers of a variable, the *smallest* power that appears in any one term is the most that can be factored out. For example, in an expression such as $a^4b + a^3c + a^2d + a^3e^4$ the smallest power of a that appears in any term is the second power — a^2. So you can factor out a^2 from all the terms because a^2 is the GCF, the greatest common factor. You can't factor anything else out of each term.

$$a^4b + a^3c + a^2d + a^3e^4 = a^2(a^2b + a^1c + d + a^1e^4)$$

TIP

Two quick checks:

- Multiply through (distribute) your answer in your head to be sure that the factored form is equivalent to the original form.

$$a^2 \times a^2 b + a^2 \times ac + a^2 \times d + a^2 \times ae^4 = a^4 b + a^3 c + a^2 d + a^3 e^4$$

- Another good way to check your work visually is to scan the terms in parentheses to make sure that they don't share the same variable.

Look at the following example of the *quick checks*. Pretend you just finished factoring the following problem.

$$x^2 y^3 + x^3 y^2 z^4 + x^4 yz = x^2 y(y^2 + x^1 y^1 z^4 + x^2 z)$$

Does your answer multiply out to become what you started with? Multiply in your head:

$$x^2 y \cdot y^2 = x^2 y^3 \quad \text{Check!}$$

$$x^2 y \cdot xyz^4 = x^3 y^2 z^4 \quad \text{Check!}$$

$$x^2 y \cdot x^2 z = x^4 yz \quad \text{Check!}$$

Those are the three terms in the original problem.

Now for the second part of the quick check. Look at what's in the parentheses of your answer. The first two terms have y and the second two have x and z, but no variable occurs in all three terms. Check!

Unlocking combinations of numbers and variables

The real test of the factoring process is combining numbers and variables, finding the GCF, and factoring successfully. Sometimes you may miss a factor or two, but a second sweep through can be done and is nothing to be ashamed of when doing algebra problems. If you do your factoring in more than one step, it really doesn't matter in what order you pull out the factors. You can do numbers first or variables first. It'll come out the same.

- Factor $12x^2 y^3 z + 18x^3 y^2 z^2 - 24xy^4 z^3$

  The GCF is $6xy^2 z$.

  So $12x^2 y^3 z + 18x^3 y^2 z^2 - 24xy^4 z^3 = 6xy^2 z(2x^1 y^1 + 3x^2 z^1 - 4y^2 z^2)$

- The greatest common factor is 100 in the following example. Even though the powers of a and b are present in the first three terms, none of them occurs in the last term. So you're out of luck.

  Factor $100a^4 b - 200a^3 b^2 + 300a^2 b^2 - 400$

So, doing the factorization,

$$100a^4 b - 200a^3 b^2 + 300a^2 b^2 - 400 = 100(a^4 b - 2a^3 b^2 + 3a^2 b^2 - 4)$$

✔ The following expression cannot be factored. It's considered prime. Even though each of the numbers is *composite* (each can be divided by values other than themselves), they have no factors in common. The three terms share nothing.

Factor $26mn^3 - 25x^2 y + 21a^4 b^4 mnxy$

✔ In this example, you see that, even if you don't divide through by the GCF the first time, all is not lost. A second run takes care of the problem. Often, doing the factorizations in two steps is easier because the numbers you're dividing through by each time are smaller, and you can do the work in your head.

Factor $484x^3 y^2 + 132x^2 y^3 - 88x^4 y^5$

Assume that you determined that the GCF of the expression in this example is $4x^2 y$.

Then $484x^3 y^2 + 132x^2 y^3 - 88x^4 y^5 =$

$4x^2 y(121x^1 y^1 + 33y^2 - 22x^2 y^4)$.

Looking at the expression in the parentheses, you can see that each of the numbers is divisible by 11 and that there's a *y* in every term. The terms in the parentheses have a GCF (greatest common factor) of 11*y*.

$4x^2 y \left[121x^1 y^1 + 33y^2 - 22x^2 y^4 \right] =$

$4x^2 y \left[11y(11x + 3y^1 - 2x^2 y^3) \right] =$

$(4x^2 y)(11y)(11x + 3y^1 - 2x^2 y^3) =$

$44x^2 y^2 (11x + 3y^1 - 2x^2 y^3)$

You can do this factorization all at the same time, using the GCF $44x^2 y^2$, but not everyone recognizes the multiples of 44 when they see them. Also, the factorization could have been done in two or more steps in a different order with different factors each time. The result always comes out the same in the end.

✔ Each term in the next example is negative; dividing out the negative from all the terms in the parentheses makes them positive.

Factor $-4ab - 8a^2 b - 12ab^2 =$

$-4ab(1 + 2a^1 + 3b^1)$

When factoring out a negative factor, be sure to change the signs of the terms.

Grouping Terms

Groups are formed when people have something in common with one another. Put 20 people on an island, leave them there for a few days, and chances are that the 20 people will form groups as they each seek out those they can relate to in some way.

The same general process can be done in factoring. The rules are a bit stricter than the preceding social situation, but the principle is the same. Find out what those algebraic principles are in this section.

When using grouping to factor:

1. **Divide the terms into groups of two terms in each.**

2. **Look for a greatest common factor (GCF) in each group of terms and factor.**

3. **Rewrite the expression in half as many terms.** (This is what happens when you factor — you get half as many terms.)

4. **Look for a GCF of the new terms.** (If there isn't a GCF of the new terms, try a different arrangement of the terms in the divisions.)

5. **Factor out the new GCF.**

Look at the following expression.

$$4xy + 4xb + ay + ab$$

You see that some terms have four in common. Some terms have x in common. And some terms have a, b, and y mixed in there, too. But all four terms do not have a variable or number in common. They can be *grouped*, though, into two parts that can be factored independently.

1. Divide the terms into groups of two terms in each.

   Group the first two terms together and then the last two.

   $$(4xy + 4xb) + (ay + ab)$$

2. Look for a GCF in each group of terms and factor.

   $$4xy + 4xb = 4x(y + b)$$

   $$ay + ab = a(y + b)$$

3. Rewrite the expression in half as many terms.

   $$4x(y + b) + a(y + b)$$

4. Look for a GCF of the new terms.

   The new GCF is $y + b$.

5. Factor out the new GCF.

$(y + b)(4x + a)$

This grouping business doesn't really help, though, unless the results of the two separate factorizations then share something. Looking at the preceding series of steps, in Step 3 each of the factored groups had $(y + b)$ in it. When this happens, the $(y + b)$ can be factored out of the newly formed terms.

$4x(y + b) + a(y + b) = (y + b)(4x + a)$

This is the factored form. If you multiply this through (distribute), then you get the four terms that you started with.

Again, the following example has nothing that all the terms share in common. But, if you group the first two and the last two, you can factor those pairs.

Factor $ax^2 y - 3a + 9x^2 y - 27$

1. Divide the terms into equal groups of two terms in each.

$(ax^2 y - 3a) + (9x^2 y - 27)$

2. Look for a GCF in each group of terms and factor.

$(ax^2 y - 3a) = a(x^2 y - 3)$
$(9x^2 y - 27) = 9(x^2 y - 3)$

3. Rewrite the expression in half as many terms.

$a(x^2 y - 3) + 9(x^2 y - 3)$

4. Look for a GCF of the new terms.

The GCF is $x^2 y - 3$

5. Factor out the new GCF.

$a(x^2 y - 3) + 9(x^2 y - 3) = (x^2 y - 3)(a + 9)$

What happens if the terms aren't in this order? How do you know what order to write them in? Do you get a different answer? Well, scramble the terms and write the problem as $ax^2 y + 9x^2 y - 27 - 3a$ and see what you have.

The first two terms have a GCF of $x^2 y$. The second two terms have a GCF of -3. Grouping and factoring gives you $x^2 y(a + 9) - 3(9 + a)$.

The expressions in the parentheses don't look exactly alike, but addition is *commutative* — you can add in either order and get the same result. You can reverse the 9 and the a in the last factor so that it looks the same as the first.

$x^2 y(a + 9) - 3(a + 9)$

Now, you can factor the $(a + 9)$ out of each term to finish the problem.

$$(a + 9)(x^2 y - 3)$$

The two factors in this answer are reversed from the first way you did the problem, but multiplication is also commutative.

In this last example, note that the two pairs of terms can be grouped and factored.

Factor $4ab^2 - 8ac^2 + 5x^2 b - 10x^2 c =$

$(4ab^2 - 8ac^2) + (5x^2 b - 10x^2 c) =$

$4a(b^2 - 2c^2) + 5x^2(b - 2c)$

The expressions in the parentheses look similar, but they aren't the same. Changing the order won't help in this case. There are now two terms, but they don't have a common factor. This expression is as simple as it can be. In other words, it's prime (in the algebraic sense).

Chapter 10

Getting the Second Degree

- -

In This Chapter

▶ Getting squared away with quadratic expressions

▶ Finding out how to FOIL without thwarting

▶ Stepping through the unFOIL process

▶ Getting organized to factor a quadratic

- -

Quadratic (second-degree) expressions, such as $3x^2 - 12$, or $-16t^2 + 32t + 11$ are studied extensively in algebra because they have so many applications in calculus and physics and other disciplines. These are *expressions* because they're made up of two or more terms with plus (+) or minus (−) signs between them. If there was an equal sign, they would be *equations*. The good news is that they're manageable. The bad news — well, there is none! Second-degree expressions are so darned nice to work with!

Quadratic expressions have a particular variable raised to the second degree. A quadratic expression can have one or more terms, and not all the terms must have a squared variable, but at least one of the terms needs to have that exponent of two. Also, a quadratic expression can't have any power greater than two. The highest power in an expression determines its name.

Some quadratic expressions may have one variable in them, such as $2x^2 - 3x + 1$. Others may have two or more variables, such as $2\pi r^2 + 2\pi rh$. They all have their place in mathematics and science. In this chapter, I show you how to make them work for you.

The Standard Quadratic Expression

The standard quadratic, or second-degree, expression has a variable that is squared, and no variables with powers higher than two in any of the terms. Where *a* is not equal to 0, quadratic expressions are of the form:

$$ax^2 + bx + c$$

You may notice that the following examples of quadratic expressions each have a variable raised to the second degree. There's no power higher than two in any of them.

$$4x^2 + 3x - 2$$

$$a^2 + 11$$

$$6y^2 - 5y$$

These expressions are usually written in terms of an *x, y, z,* or *w.* The letters at the end of the alphabet are used more frequently for the variable, while those at the beginning of the alphabet are usually used for a number or constant. This doesn't *have* to be the case, but it is *usually* the case.

In a quadratic expression, the *a* — the variable raised to the second power — can't be zero. If *a* was allowed to be 0, then the x^2 would be multiplied by zero. It wouldn't be a quadratic expression any more. The variables *b* or *c* can be 0, but *a* cannot.

Quadratics don't necessarily have all positive terms, either. The standard form, $ax^2 + bx + c$, is written with all positives for convenience. But if *a, b,* or *c* represents a negative number, then that term would be negative.

The terms are usually written with the second-degree term first, the first-degree term next, and the number last.

Quadratic expressions you know

You're probably already familiar with a few of these quadratic expressions and equations:

✔ $E = mc^2$: Einstein said it first! This equation has only the one term because that's all old Al needed. The E stands for energy, *m* for mass, and the *c* is a constant representing the speed of light.

✔ $A = s^2$: You may have used this equation back in grade school. The area of a square is found by taking the length of a side and raising that number to the second degree. The *A* stands for the area and the *s* for the length of the side.

✔ $a^2 + b^2 = c^2$: This is Pythagoras' famous contribution to the understanding of the right triangle. *a* and *b* represent the lengths of the two shorter sides in a right triangle, and *c* represents the length of the hypotenuse — the longest side.

✔ $A = \pi r^2$: This is used to find the area of a circle. There's the argument that "pi are squared" is wrong, because "pi are *round*" — sorry, couldn't resist that one.

These are just a few of the more famous quadratic expressions and equations. When you come across them in real-life problems you may have to factor them on your way to solving equations or answering questions involving them.

TIP

If an expression has more than one variable, decide which variable makes it a quadratic expression (look for the variable that's squared) and write the expression in terms of that variable. This means, after you find the variable that's squared, write the rest of the expression in decreasing powers of that variable.

$$aby + cdy^2 + ef$$

This can be a second-degree expression in y. Written in the standard form for quadratics, $ax^2 + bx + c$, where the second degree term comes first, it looks like

$$(cd)y^2 + (ab)y + ef$$

The parentheses aren't necessary in the preceding case and doesn't change anything, but they're used sometimes for emphasis. The parentheses just make seeing the different parts easier.

Here's an example where you get to make choices.

$$a^2bx + cdx^2 + aef$$

This can be a second-degree expression in terms of a or x.

Second degree in a: $(bx)a^2 + (ef)a + cdx^2$

Even though there's a second-degree factor of x in the last term, that term is thought of as a constant, a value that doesn't change, rather than a variable if the expression is a to the second degree.

Now, changing roles, the second degree in x: $(cd)x^2 + (a^2b)x + aef$.

Reining in Big Numbers

Some perfectly good quadratic expressions are just too awkward to handle. Some can be made better by factoring. Some are just going to be uncooperative — you're stuck with them. In this section, I go back to my favorite standby: finding a greatest common factor. If the terms in the quadratic have something in common, then that can be factored out, leaving something more reasonable to deal with.

✔ The following quadratic operation can be made more useable by factoring out the common factor and arranging the result in a nice, organized expression.

$$800x^2 + 40,000x - 100,000$$

It has large numbers, but each number can be evenly divided by 800 — a common factor.

$$800x^2 + 40{,}000x - 100{,}000 =$$

$$800(x^2 + 50x - 125)$$

✔ The following quadratic operation has four different variables with powers of two. Only the x, though, appears in a term with a power of one. So, you may choose to write this as a quadratic in x and factor out some of the other variables.

$$a^2x^2 + a^2c^2 + a^2b^2x$$

Rewrite the expression in decreasing powers of x.

$$a^2x^2 + a^2b^2x + a^2c^2$$

Find the GCF, which is a^2, and factor it out.

$$a^2(x^2 + b^2x + c^2)$$

FOILing

What is FOIL?

a) A brainchild of Mr. Reynolds Aluminum.

b) An expression of dismay: "Rats! FOILed again!"

c) An acronym for *first, outer, inner,* and *last.*

Choice C is my final answer. FOIL is an acronym that cropped up somewhere between my high school years and my teaching years. It was sort of "under the counter" at first. Respected mathematicians didn't want to use it. But it has caught on and is now accepted, published, and used extensively in the algebra classroom. FOIL is easy to remember and apply.

Algebra-speak

Just as a *phrase,* such as *a hill of beans,* lacks a verb in the English language, an algebraic *expression,* such as $6x^2 + 11$, lacks an equal sign. Neither the phrase nor the algebraic expression makes any assertions.

On the other hand, a *statement* or an algebraic *equation* makes an assertion. A statement must contain a verb, which is similar to the equal sign in an algebraic equation. For example, you may say that $6 + 3$ *is* 9. A statement, such as "The car is worth \$15,000," can be true or false depending on what car you are referring to. A mathematical statement, or an equation, such as $6x - 1y = 11$, makes a claim. Whether the claim is true depends on what the x and y are.

This chapter is on factoring, but first you need to find out how to multiply two binomials together using FOIL. Chapter 8 shows you how to multiply two binomials together by distributing. This chapter gives you an alternate method.

FOILing basics

Many quadratic expressions, such as $6x^2 + 7x - 3$, are the result of multiplying two binomials (two terms separated by addition or subtraction) together, so you can undo the multiplication by factoring them.

$$6x^2 + 7x - 3 = (2x + 3)(3x - 1)$$

The right side is the *factored* form. But how can you tell that the left side of that equation is equal to the right side just looking at it? It's not like a greatest common factor, when you look for something in common.

What does FOIL stand for? The letters each refer to two terms — one from each of two binomials — multiplied together in a certain order. The steps don't *have* to be done in this order, but they usually are. Otherwise, the acronym would be something like OFIL (heaven forbid). The following list describes what each letter in the FOIL acronym stands for:

- **F** stands for the *first* term in each binomial: $(\mathbf{3a} + 6)(\mathbf{2a} - 1)$
- **O** stands for the two *outermost* terms — those farthest to the left and right: $(\mathbf{3a} + 6)(2a - \mathbf{1})$
- **I** stands for the *inner* terms in the middle: $(3a + \mathbf{6})(\mathbf{2a} - 1)$
- **L** stands for the *last* term in each binomial: $(3a + \mathbf{6})(2a - \mathbf{1})$

In each binomial, there's the left term and the right term. But the two terms have other names, also. (Just like someone named Michael may be "Mike" to one person and "son" to another.) The other names for the terms in the binomials refer to their positions in terms of the whole picture. The two terms not in the middle are the *outer* terms. The two terms in the middle are the *inner* terms. Use this as an example: $(a + b)(c + d)$. The terms a and c are *first;* the terms b and d are *last* in each binomial. The terms a and d are *outer;* the terms b and c are *inner* in the big picture. As you can see, each term has two names. In the problem $(2x + 3)(3x - 1)$ the term $2x$ is called *first* one time and *outer* another time. That's okay.

Figure 10-1 gives you a visual on how this is done.

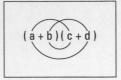

FOILed again, and again

The following steps demonstrate how to use FOIL on a multiplication problem: $(2x + 3)(3x - 1)$.

1. **Multiply the *first* term of each binomial together.**

    $(2x + 3)(3x - 1)$

    $(2x)(3x) = 6x^2$

2. **Multiply the *outer* terms together.**

    $(2x + 3)(3x - 1)$

    $(2x)(-1) = -2x$

3. **Multiply the *inner* terms together.**

    $(2x + 3)(3x - 1)$

    $(3)(3x) = 9x$

4. **Multiply the *last* term of each expression together.**

    $(2x + 3)(3x - 1)$

    $(3)(-1) = -3$

5. **List the four results of FOIL in order.**

    $6x^2 - 2x + 9x - 3$

6. **Combine the like terms.**

    $6x^2 + 7x - 3$

Distributing the two terms in the first binomial over the second produces the same result, but in the case of binomials, using FOIL is easier. For more on distributing, see Chapter 8.

See how the FOIL numbered steps work on a couple of negative terms in the following example.

   $(x - 3)(2x - 9)$

1. Multiply the first terms.

    $(x)(2x) = 2x^2$

2. Multiply the outer terms.

$$(x)(-9) = -9x$$

3. Multiply the inner terms.

$$(-3)(2x) = -6x$$

4. Multiply the last terms.

$$(-3)(-9) = 27$$

5. List the four results of FOIL in order.

$$2x^2 - 9x - 6x + 27$$

6. Combine the like terms.

$$2x^2 - 15x + 27$$

The following example is a bit more complicated to do, but FOIL makes it much easier. The tasks are broken down into smaller, simpler tasks, and then the results are combined for the final result.

The next steps take you through $[x + (y - 4)][3x + (2y + 1)]$.

1. Multiply the first terms.

$$(x)(3x) = 3x^2$$

2. Multiply the outer terms.

$$(x)(2y + 1) = 2xy + x$$

3. Multiply the inner terms.

$$(y - 4)(3x) = 3xy - 12x$$

4. Multiply the last terms. The last terms are two binomials, too. You FOIL these binomials when you finish this series of FOIL steps.

$$(y - 4)(2y + 1)$$

5. List the four results of FOIL in order.

$$3x^2 + 2xy + x + 3xy - 12x + (y - 4)(2y + 1)$$

6. Combine like terms.

$$3x^2 + 5xy - 11x + (y - 4)(2y + 1)$$

Notice the product of two binomials from the preceding Step 4: $(y - 4)(2y + 1)$. You can FOIL them.

1. Multiply the first terms: $(y)(2y) = 2y^2$

2. Multiply the outer terms: $(y)(1) = y$

3. Multiply the inner terms: $(-4)(2y) = -8y$

4. Multiply the last terms: $(-4)(1) = -4$

5. Write the results in order: $2y^2 + y - 8y - 4$

6. Combine like terms: $2y^2 - 7y - 4$

   Now, replace the two binomials multiplied together with this new result, and you can rewrite the entire problem.

   $3x^2 + 5xy - 11x + 2y^2 - 7y - 4$

This may seem complicated, but using FOIL is easier than doing all the distributing. Do you remember the rule for multiplying the sum of any two terms by their difference? If your answer is "Yes," then skip this problem, give yourself a pat on the back, and move to the head of the class. If your answer is "No," then just give yourself a pat on the back and plod on.

The sum of any binomial multiplied by the difference of the same two terms (see the operation that follows) is an easy operation because the middle terms cancel each other out — they both have the same absolute value, except one is positive and the other is negative.

$$(a + b)(a - b) = a(a - b) + b(a - b) = a^2 - ab + ab - b^2 = a^2 - b^2$$

The following operation multiplies the sum and difference of the same two values. In Chapter 8, I show you how the middle terms cancel each other out or disappear. This is even more evident with FOIL.

$(5x - 3)(5x + 3)$

1. Multiply the first terms: $(5x)(5x) = (5x)^2 = 25x^2$

2. Multiply the outer terms: $(5x)(3) = 15x$

3. Multiply the inner terms: $(-3)(5x) = -15x$

4. Multiply the last terms: $(-3)(+ 3) = -9$

5. Write the results in order: $25x^2 + 15x - 15x - 9$

6. Combine like terms: The products, $15x$ and $-15x$, are opposites of each other. The first and last products are all that's left.

   $25x^2 - 9$

Take a look at the quadratic operations that follow. Are they good examples of the sum and difference of binomials and FOIL?

$$(3x + 2)(3x - 2) = 9x^2 - 4$$
$$(2z - m)(2z + m) = 4z^2 - m^2$$
$$(m^2 - n^2)(m^2 + n^2) = m^4 - n^4$$

FUN FACT

Carl Friedrich Gauss, child prodigy

The mathematician Carl Friedrich Gauss, a child prodigy, was only three years old when he corrected some calculations in his father's payroll records. He went on to make significant contributions to mathematics.

Legend has it that when Gauss was a schoolboy, his tired teacher told the class to find the sum of the numbers from 1 through 100 to keep them occupied so he could rest. Moments later, little Carl Friedrich was at the teacher's elbow with a solution. The teacher looked in disbelief at the boy's answer, which, of course, was correct. Gauss wasn't a whiz at adding. He just

got organized and found patterns in the numbers to make the adding easier and much more interesting. He saw that $1 + 99 = 100$, $2 + 98 = 100$, $3 + 97 = 100$, and so on. The sum of 49 of these 100s, the 50 in the middle, and the 100 at the end is 5,050.

Thanks to Gauss, a standard formula is available for the sum of any list of consecutive integers. This formula is $S = n(n + 1) \div 2$. The S represents the sum of the numbers, and the n is the biggest, or last, number in the list that starts with the number one.

UnFOILing

When you look at an expression such as $2x^2 - 5x - 12$, you may think that figuring out how to factor this into the product of two binomials is an awful chore. And you may wonder whether it even can be factored that way. Let me assure you that these problems are really quite easy.

The nice thing is that there's a system to make unFOILing simple. You go through the system, and it helps you find what the answer is or helps you determine if there *isn't* an answer. This can't be said about all factoring problems, but it is true of quadratics in the form $ax^2 + bx + c$. That's why quadratics are so nice to work with in algebra.

The key to unFOILing these factoring problems is being organized.

- ✔ Be sure you have an expression in the form $ax^2 + bx + c$.
- ✔ Be sure the terms are written in the order of decreasing powers.
- ✔ If needed, review the lists of prime numbers and perfect squares (see the Cheat Sheet for some of both).
- ✔ Follow the steps.

ALGEBRA RULES
1
+1
2

Follow these steps to factor the quadratic $ax^2 + bx + c$, which is written in the order of decreasing powers, using unFOIL.

1. **Determine all the ways you can multiply two numbers to get a.**

   Every number can be written as at least one product, even if it's only the number times one. So assume that there are two numbers, e and f, whose product is equal to a: $a = e \cdot f$. These are the two numbers you want for this problem.

2. **Determine all the ways you can multiply two numbers together to get c.**

   If the value of c is negative, ignore the negative sign for the moment. Concentrate on what factors result in the absolute value of c.

   Now assume that there are two numbers, g and h, whose product is equal to c: $c = g \cdot h$. Use these two numbers for this problem.

3. **Now look at the sign of c and your lists from Steps 1 and 2.**

   1. If c is positive, find a value from your Step 1 list and another from your Step 2 list such that the sum of their product and the product of the two remaining numbers in those steps results in b.

      Assume, using $e \cdot f$ and $g \cdot h$, that $e \cdot g + f \cdot h = b$.

   2. If c is negative, find a value from your Step 1 list and another from your Step 2 list such that the difference of their product and the product of two remaining numbers from those steps results in b.

      Assume, using $e \cdot f$ and $g \cdot h$, that $e \cdot g - f \cdot h = b$.

4. **Arrange your choices as binomials.**

   The e and f have to be in the *first* positions in the binomials, and the g and h have to be in the *last* positions. They have to be arranged so the multiplications in Step 3 have the correct *outer* and *inner* products.

   $(e \quad h)(f \quad g)$

5. **Place the signs appropriately.**

   The signs are both positive if c is positive and b is positive.

   The signs are both negative if c is positive and b is negative.

   One sign is positive and one negative if c is negative; the choice depends on whether b is positive or negative and how you arranged the factors.

Using unFOIL, follow these steps to factor the quadratic $2x^2 - 5x - 12$, which is in the form $ax^2 + bx + c$ and written in the order of decreasing powers.

1. Determine all the ways you can multiply two numbers to get a.

   You can get these numbers from the prime factorization of a. See Chapter 7 if you need to review prime factorizations.

   Prime numbers can only be divided by themselves and one. Prime factorization is finding the prime numbers that divide any given value.

Sometimes, writing out the list of ways to multiply is a big help.

In the example $2x^2 - 5x - 12$, the value of a is 2. The only way to multiply two numbers together to get 2 is 1×2.

2. Determine all the ways you can multiply two numbers to get c.

Again, referring to the example, $2x^2 - 5x - 12$, the value of c is –12. Ignore the negative sign right now. The negative becomes important in the next step. Just concentrate on what multiplies together to give you 12.

There are three ways to multiply two numbers together to get 12: 1×12, 2×6, and 3×4.

3. Look at the sign of c and your lists from Steps 1 and 2.

1. If c is positive, find a value from your Step 1 list and another from your Step 2 list such that the sum of their product and the product of the two remaining numbers in those steps results in b.

2. If c is negative, find a value from your Step 1 list and another from your Step 2 list such that the difference of their product and the product of two remaining numbers from those steps results in b.

4. Choose a product from Step 1 and a product from Step 2.

In the case of the example, c is –12 and b is 5. So, look for a combination from Step 1 and Step 2 whose difference results in 5.

Use the 1×2 from Step 1 and the 3×4 from Step 2. Multiply the 1 from Step 1 times the 3 from Step 2 and then multiply the 2 from Step 1 times the 4 from Step 2.

$(1)(3) = 3$ and $(2)(4) = 8$

The two products are 3 and 8, whose difference is 5.

$8 - 3 = 5.$

5. Arrange your choices as binomials so the results are those you want.

From the example, the following arrangement multiplies the $(1x)(2x)$ to get the $2x^2$ needed for the first product. Likewise, the 4 and 3 multiply to give you 12. The outer product is $3x$ and the inner product is $8x$.

$(1x \quad 4)(2x \quad 3)$

6. Place the signs to give the desired results.

$(1x - 4)(2x + 3) = 2x^2 - 5x - 12$

This next example, $24x^2 - 34x - 45$, offers many numbers to choose from.

1. Determine all the ways you can multiply two numbers to get a.

a is 24, which equals 1×24, 2×12, 3×8, or 4×6.

2. Determine all the ways you can multiply two numbers to get c.

c is 45, which equals 1×45, 3×15, or 5×9.

3. Look at the sign of c and your lists from Steps 1 and 2 to see if you want a *sum* or *difference*.

   c is negative, so you want a *difference* of 34 between products.

4. Choose a product from Step 1 and a product from Step 2.

   Use the 4×6 from *a* and the 5×9 from *c*. The product of 4 and 5 is 20. The product of 6 and 9 is 54. The difference of these products is 34.

5. Arrange your choices as binomials so the results are those you want.

   $(4x \quad 9)(6x \quad 5)$ line up products the way you want.

6. Place the signs to give the desired results.

   $(4x - 9)(6x + 5) = 24x^2 - 34x - 45$

The combinations you want may not just leap out at you. But having a list of all the possibilities helps heaps. You can start systematically trying out the different combinations. For instance, take the 1×24 and try it with all three sets of numbers that give you *c*: 1×45, 3×15, or 5×9. If none of those work, then try the 2×12 with all the sets of numbers that give you *c*. Continue until you've systematically gone through all the possible combinations. If none works, you know you're done. Doing it this way doesn't leave you wondering if you've missed anything.

In the next example, $2x^2 - 9x + 4$, the sum of outer and inner is used.

1. Determine all the ways you can multiply two numbers to get *a*.

   a is 2, which can only be 1×2.

2. Determine all the ways you can multiply two numbers to get *c*.

   c is 4, which can be written as 1×4 or 2×2.

3. Look at the sign of c and your lists from Steps 1 and 2 to see if you want a sum or difference.

   c is positive, so you want the *sum* of the outer and inner products.

4. Choose a product from Step 1 and a product from Step 2.

   Using the 1×2 and the 1×4 factors, multiply $(2)(4)$ to get 8, and multiply the two ones together to get 1. The sum of the 8 and the 1 is 9.

5. Arrange your choices as binomials so the results are those you want.

   $(2x \quad 1)(1x \quad 4)$

6. Place the signs to give the desired results.

   $(2x - 1)(1x - 4) = 2x^2 - 9x + 4$

In the next example, $10x^2 + 31x + 15$, all the terms are positive. The sum of the outer and inner products will be used. And there are several choices for the multipliers.

1. Determine all the ways you can multiply two numbers to get a.

   The 10 can be written as 1×10 or 2×5.

2. Determine all the ways you can multiply two numbers to get c.

   The 15 can be written as 1×15 or 3×5.

3. Look at the sign of c and your lists from Steps 1 and 2 to see if you want a sum or difference.

   The last term is positive, so you want the *sum* of the products to be 31.

4. Choose a product from Step 1 and a product from Step 2.

   Using the 2×5 and the 3×5, multiply $(2)(3)$ to get 6, and multiply $(5)(5)$ to get 25. The sum of 6 and 25 is 31.

5. Arrange your choices in the binomials so the results are those you want.

   $(2x \quad 5)(5x \quad 3)$

6. Place the signs to give the desired results.

   $(2x + 5)(5x + 3) = 10x^2 + 31x + 15$

This last example, $18x^2 - 27x - 4$, looks, at first, like a great candidate for factoring by this method. You'll see, though, that not everything can factor. Also, the point can be made that using this method assures you that you've "left no stone unturned."

1. Determine all the ways you can multiply two numbers to get a.

   The 18 can be written as 1×18 or 2×9 or 3×6.

2. Determine all the ways you can multiply two numbers to get c.

   The 4 can be written as 1×4 or 2×2.

3. Look at the sign of c and your lists from Steps 1 and 2 to see if you want a sum or difference.

   The last term is negative, so you want the *difference* of the products to be 27.

4. Choose a product from Step 1 and a product from Step 2.

   You can't seem to find any combination that gives you a difference of 27. Run through them to be sure that you haven't missed anything.

Using the 1×18, cross it with:

1×4 gives you a difference of either 14, using the (1)(4) and (18)(1) or 71 using the (1)(1) and the (18)(4).

2×2 gives you a difference of 34 using (1)(2) and (18)(2); there's only one choice because both of the second factors are 2.

Using the 2×9, cross it with:

1×4 gives you a difference of either 34 using (2)(1) and (9)(4) or 1, using (2)(4) and (9)(1).

2×2 gives you a difference of 14, only.

Using the 3×6, cross it with:

1×4 gives you a difference of either 21, using (3)(1) and (6)(4) or 6, using (3)(4) and (6)(1).

2×2 gives you a difference of 6, only.

Because you exhausted all the possibilities and have not been able to create a difference of 27, you can assume that this quadratic cannot be factored. It is *prime*.

Factoring Several Expressions

Sometimes you have to factor a problem more than once. This section shows you how two completely different factoring techniques can be used on the same problem. The process of using different factoring techniques is different from reusing the same methods, such as taking out common factors several times.

A quadratic, such as $40x^2 - 40x - 240$, can be factored using two different techniques, which can be done in two different orders, making the problem *easier* and *harder*. It's the order in which the factoring is done that makes one way easier and the other way harder.

I'll show you the harder method first, so you'll see why it's important to make a good choice. In this case, the big numbers are left in and the unFOILing is done first.

1. **Determine all the ways you can multiply two numbers to get *a*.**

   The 40 can be written as 1×4, 2×2, 4×10, or 5×8.

2. **Determine all the ways you can multiply two numbers to get *c*.**

   The 240 can be written as 1×240, 2×120, 3×80, 4×60, 5×48, 6×40, 8×30, 10×24, 12×20, or 15×18.

3. **Look at the sign of *c* and your lists from Steps 1 and 2.**

   The last term is negative, so you want the *difference* of the products to be 40.

   Use the 4×10 and the 12×20, multiply $(4)(20)$ to get 80, and multiply $(12)(10)$ to get 120. The difference between 80 and 120 is 40.

4. **Arrange your choices as binomials and place the signs appropriately.**

   $$(4x - 12)(10x + 20) = 40x^2 - 40x - 240$$

   But just look at those binomials. Each of them can be factored themselves. The terms in the first binomial can each be factored by 4, and the terms in the second binomial can each be factored by 10.

   $$(4x - 12)(10x + 20) = 4(x - 3)10(x + 2) = 40(x - 3)(x + 2)$$

Next, try the easier way. Factor out the greatest common factor (GCF), first.

$$40x^2 - 40x - 240 = 40(x^2 - x - 6)$$

1. **Just look inside the parenthesis.**

   The 1 can be written only as 1×1. The 6 can be written as 1×6 or 2×3. Notice how the list of choices is much shorter and more manageable than if you try to unFOIL before factoring out the GCF.

2. **Looking at the sign of c, choose your products.**

   The last term is negative, so you want the *difference* of the products to be 1. Using the 1×1 and the 2×3, it's easy to set up the factors:

   $$(1x \quad 2)(1x \quad 3)$$

   The middle term, x, is negative, so you want the $3x$, the *outer* terms, to be negative. Put the 40 that you factored out in the first place back into the answer.

   $$40x^2 - 40x - 240 = 40(x - 2)(x - 3)$$

You can get to the correct answer no matter what you choose to do in what order. As a general rule, though, it's usually best to factor out a GCF first.

Chapter 11

Factoring Special Cases

· ·

In This Chapter

▶ Paring down perfect cubes

▶ Sorting out the difference of two squares

▶ Summing up two perfect cubes

▶ Putting polynomials with several terms in order

· ·

*T*his chapter has some mighty helpful factoring information that doesn't belong under linear or quadratic factoring rules. You may want to look at Chapters 9 and 10 for more factoring rules and tips. Half of the factoring process is knowing the rules and the other half is recognizing when to use what rule. These are equally important skills — you need both.

Befitting Binomials

If a binomial (two-term) expression can be factored at all, it must be factored in one of four ways. First, look at the addition or subtraction sign that always separates the two terms within a binomial. Then look at the two terms. Are they squares? Are they cubes? Are they nothing special at all? The nice thing about having two terms in an expression is that you have four and only four ways to check.

The four ways to factor a binomial are

✔ Finding the greatest common factor (GCF)

✔ Factoring the difference of two perfect squares

✔ Factoring the difference of two perfect cubes

✔ Factoring the sum of two perfect cubes

When you have a factoring problem with two terms, you can go through the list to see which way works. Sometimes the two terms can be factored in more than one way, such as finding the GCF *and* the difference of two squares. After you go through one factoring method, check inside the parenthesis to see if another factoring can be done. If you checked each item on the list of ways to factor and none works, then you know that the expression *can't* be factored any further. You can stop looking and say you're done.

Finding the greatest common factor (GCF) is always an easy and quick option to look into when factoring: (For more on how to find the GCF, see Chapter 9.) What's left after factoring is much easier to deal with. But do read the following sections to discover other factoring pearls of wisdom.

Factoring the difference of two perfect squares

If two terms in a binomial are perfect squares and they're separated by subtraction, then they can be factored. A perfect square is *not* a reference to that ol' high school prom date with two left feet who refused to dance the entire evening. A *perfect square* is the result of multiplying another number by itself. Twenty-five is a perfect square because it's equal to five times five. To factor one of these binomials, just find the square roots of the two terms that are perfect squares and write the factorization as the sum and difference of the square roots. For example, $x^2 - y^2 = (x + y)(x - y)$. However, looking at another example, the binomial $x^2 + 3$ is *not* the difference of two perfect squares because there's a plus sign, not a minus, and 3 isn't a perfect square.

If subtraction separates two squared terms, then the sum and difference of the two square roots factors the binomial.

$$a^2 - b^2 = (a + b)(a - b)$$

✔ Factor $9x^2 - 16$. The square roots of $9x^2$ and 16 are $3x$ and 4, respectively. The sum of the roots is $3x + 4$ and the difference between the roots is $3x - 4$. So, $9x^2 - 16 = (3x + 4)(3x - 4)$.

✔ Factor $25z^2 - 81y^2$. The square roots of $25z^2$ and $81y^2$ are $5z$ and $9y$, respectively. So, $25z^2 - 81y^2 = (5z + 9y)(5z - 9y)$.

✔ Factor $x^4 - y^6$. The square roots of x^4 and y^6 are x^2 and y^3, respectively. So $x^4 - y^6 = (x^2 + y^3)(x^2 - y^3)$.

✔ Factor $x^2 - 3$. In this case, the second number is not a perfect square. But sometimes it's preferable to have the expression factored, anyway. The square root of x^2 is x, and you can write the square root of 3 as $\sqrt{3}$. (For more on square roots and radicals, see Chapter 4.) Now the factorization can be written: $x^2 - 3 = (x + \sqrt{3})(x - \sqrt{3})$.

Factoring the difference of perfect cubes

A *perfect cube* is the number you get when you multiply a number times itself then multiply the answer times the first number again. A cube is the third power of a number. The difference of two cubes is a binomial expression $a^3 - b^3$.

The most well-known perfect cubes are those whose roots are integers, not decimals. Check the Cheat Sheet for a list of the first 10 positive integers cubed. Becoming familiar with and recognizing these cubes in an algebra problem can save you time and improve your accuracy.

Refer to Chapter 4 if these rules for working with exponents don't ring a bell. In the first example, the cube outside the parenthesis means each variable gets raised to that power. In the second example, the rule involving raising a power to a power is used. Both results are cubes.

$$(yz)^3 = y^3 \cdot z^3$$
$$(a^2)^3 = a^6$$

Variable cubes are relatively easy to spot because their exponents are always divisible by three. When a number is cubed and multiplied out, you can't always tell it's a cube unless you memorize the cubes or refer to lists.

Look at the following list. These expressions are the difference of cubes that *can* be factored. Each term is a cube — the numbers all have a cube root (the number that's multiplied by itself three times) most of which are on the list of cubes. The variables all have powers that are multiples of three.

- ✔ $m^3 - 8$
- ✔ $1,000 - 27z^3$
- ✔ $64x^6 - 125y^{15}$

Notice that every number is a perfect cube and that every variable has a power that is a multiple of three.

To factor the difference of two perfect cubes, remember that the difference of two perfect cubes equals the difference of their cube roots multiplied by the sum of their squares and the product of their cube roots.

$$a^3 - b^3 = (a - b)(a^2 + ab + b^2)$$

The results of factoring the difference of perfect cubes are:

✔ A binomial factor $(a - b)$ made up of the two cube roots of the perfect cubes separated by a minus sign. To find the cube root, refer to the list of cubes on the Cheat Sheet. If the cube isn't there, and the number is smaller than the largest cube on the list, then the number isn't a perfect cube. For bigger numbers, use a scientific calculator and the cube root button.

✔ A trinomial factor $(a^2 + ab + b^2)$ made up of the squares of the two cube roots added to the product of the cube roots in the middle. Remember, a trinomial has three terms, and this one has all plus signs in it.

The following three examples show you how the rule works. The first example isn't one you'd usually see in algebra, but I included it to convince any Doubting Thomas.

✔ In this example, use only numbers. Factor the difference between the cubes 216 and 125 so that you can see how this rule really works.

$$216 - 125$$

Use the rule $a^3 - b^3 = (a - b)(a^2 + ab + b^2)$, so that 216 is the a^3; 125 is the b^3.

The cube root of 216 is 6; the cube root of 125 is 5, so 6 is the a; 5 is the b.

Also, 36 is a^2, 25 is b^2, and 30 is the product ab.

Substituting into $a^3 - b^3 = (a - b)(a^2 + ab + b^2)$, you get $216 - 125 = (6 - 5)(36 + 30 + 25)$.

Check to see if the equation is true. The difference between 216 and 125 is 91, and $36 + 30 + 25$ equals 91.

$$91 = (1)(91)$$

This shows that whether the expression is the difference of the two cubes or the factored form, the answer comes out the same.

This doesn't really *prove* anything, but it's a nice demonstration that the method works on numbers.

✔ Factor $m^3 - 8$.

The cube root of m^3 is m, and the cube root of 8 is 2.

$$m^3 - 8 = (m - 2)(m^2 + 2m + 4)$$

Notice that the sign between the m and the 2 is the same as the sign between the cubes. The square of m is m^2 and the square of 2 is 4. The product of the two cube roots is $2m$, and the signs in the trinomial are all positive.

✔ Factor $64x^3 - 27y^6$.

The cube root of $64x^3$ is $4x$, and the cube root of $27y^6$ is $3y^2$.

The square of $4x$ is $16x^2$, the square of $3y^2$ is $(3y^2)^2 = 9y^4$, and the product of $(4x)(3y^2)$ is $12xy^2$.

$$64x^3 - 27y^6 = (4x - 3y^2)(16x^2 + 12xy^2 + 9y^4)$$

✔ Factor $a^3b^6c^9 - 1331d^{300}$.

The cube root of $a^3b^6c^9$ is ab^2c^3, and the cube root of $1331d^{300}$ is $11d^{100}$.

The square of ab^2c^3 is $a^2b^4c^6$, and the square of $11d^{100}$ is $121d^{200}$. The product of $(ab^2c^3)(11d^{100})$ is $11ab^2c^3d^{100}$.

$$a^3b^6c^9 - 1331d^{300} = (ab^2c^3 - 11d^{100})(a^2b^4c^6 + 11ab^2c^3d^{100} + 121d^{200})$$

Factoring the sum of perfect cubes

You have a break coming. The rule for factoring the sum of two perfect cubes is almost the same as the rule for factoring the difference between perfect cubes, which I cover in the previous section. You just have to change two little signs to make it work.

The sum of two cubes equals the sum of its roots times the squares of its roots minus the product of the roots.

$$a^3 + b^3 = (a + b)(a^2 - ab + b^2)$$

Like the results of factoring the difference of two cubes, the results of factoring the sum of two cubes is also made up of a binomial factor $(a + b)$ and a trinomial factor $(a^2 - ab + b^2)$.

Notice that the sign between the two cube roots $(a + b)$ is the same as the sign in the problem to be factored $(a^3 + b^3)$. The squares in the trinomial expression are still both positive, but you change the sign of the middle term to minus.

✔ Look at this practical example to get a clearer idea of how to factor the sum of two cubes.

$$1,000z^3 + 343$$

The cube root of $1,000z^3$ is $10z$, and the cube root of 343 is 7. The product of $10z$ and 7 is $70z$.

$$1,000z^3 + 343 = (10z + 7)(100z^2 - 70z + 49)$$

Great leaders make great mathematicians

Two famous leaders, Napoleon Bonaparte and U. S. President James Garfield, were drawn to the mysteries of mathematics. Napoleon Bonaparte fancied himself an amateur geometer and liked to hang out with mathematicians — they're so much fun!

The Napoleon theorem, which he named for himself, says that if you take any triangle and construct equilateral triangles on each of the three sides and find the center of each of these three triangles and connect them, the connecting segments always form another equilateral triangle. Not bad for someone who met his Waterloo!

The twentieth U.S. President, James Garfield, also dabbled in mathematics and discovered a new proof for the Pythagorean theorem, which is done with a trapezoid consisting of three right triangles and working with the areas of the triangles.

Tinkering with Trinomials and More

You can choose from a total of two methods to factor expressions with three terms, also known as *trinomials:*

- Finding the GCF
- UnFOILing

For more information on both methods, see Chapter 9. Any factoring problem is a matter of recognizing what you have so you know what method to apply. With trinomials, you can use unFOIL if the trinomial is of the form $ax^2 + bx + c$. You can find the GCF if a common factor is available. After checking for both of these situations, if neither fits, then you're done! The trinomial can't be factored.

Basically, you can factor an expression with four, six, eight, or more terms either by finding the GCF or by grouping. Chapter 9 covers finding the greatest common factor in detail, and grouping polynomials is covered in the next section.

Grouping

The other choice for factoring four or more terms is to try grouping terms to make the new groups factorable. Grouping is first covered in Chapter 9, but this chapter broadens the process to more terms and other types of grouping.

Splitting four terms into two groups of two terms

The most common form of grouping is finding a different common factor in each of the two groups formed from four terms. First, split the four terms down the middle to start forming groups. If this doesn't work, put the terms with something in common in the same group to change the order of the terms. You take the common factor out of each group, separately, and then hope to find a *new* common factor in the new terms. The new common factor usually results from factoring each separate group. The following example may help you to understand grouping four or more terms:

✔ Factor $8ax + 12ay + 10bx + 15by$.

The first two terms share the common factor $4a$, and the second two terms have a common factor of $5b$.

$$8ax + 12ay + 10bx + 15by = 4a\,(2x + 3y) + 5b(2x + 3y)$$

In this case, the grouping resulted in two new terms, each with a factor of $(2x + 3y)$. This new factor is common to both, so the GCF method is going to work.

Take that common factor out of the two new terms and see what you have.

$$4a(2x + 3y) + 5b(2x + 3y) = (4a + 5b)\,(2x + 3y)$$

Done!

Dividing up six terms into two or three groups

The next example has six terms. Because two and three divide six evenly, there's the chance that the groups can be two groups of three terms or three groups of two terms each. Sometimes you can do it either way. Sometimes you just have to try until you find the right way. In this case, both ways work.

✔ Factor by dividing the terms into two *groups of three*.

$$ax^2 + 3ax + 2a + bx^2 + 3bx + 2b$$

The first three terms have a common factor of a, and the second three terms have a common factor of b.

$$a(x^2 + 3x + 2) + b(x^2 + 3x + 2)$$

There are now two groups, each with a common factor of $(x^2 + 3x + 2)$.

$$a(x^2 + 3x + 2) + b(x^2 + 3x + 2) = (x^2 + 3x + 2)(a + b)$$

Notice that the first factor is a quadratic that can be factored with unFOIL.

$$(x^2 + 3x + 2) = (x + 1)(x + 2)$$

$$(x^2 + 3x + 2)(a + b) = (x + 1)(x + 2)(a + b)$$

✔ Grouping the terms *two at a time* is another way to work this problem. Rearrange the terms, putting the x^2 variables, the x variables, and the numbers together.

$$ax^2 + bx^2 + 3ax + 3bx + 2a + 2b$$

The first two terms have a common factor of x^2, the third and fourth terms have a common factor of $3x$, and the last two terms have a common factor of 2.

$$x^2 (a + b) + 3x(a + b) + 2(a + b)$$

Now you have three terms, each with a factor of $(a + b)$. Take the $(a + b)$ out of each term to get:

$$(a + b)(x^2 + 3x + 2) = (a + b)(x + 1)(x + 2)$$

Which way do you like best? You can choose whichever you find easiest and be assured that you'll get the same answer as everyone else.

Uneven grouping

The grouping so far has involved four or six terms divided into equal-size groups. Sometimes four terms can be separated into unequal groupings with three terms in one group and one term in the other. The way to spot these is to look for squares. Of course, you usually don't even look for unequal groupings unless other grouping methods have failed you.

✔ Factor $x^2 + 8x + 16 - y^2$.

This has four terms, but there's no good equal pairing of terms that will give you a set of useful common factors. Another option is to group *unevenly*. Group the first three terms together because they form a trinomial that can be factored. That leaves the last term by itself.

$$x^2 + 8x + 16 - y^2 = (x^2 + 8x + 16) - y^2$$

Factor the trinomial using unFOIL.

$$(x + 4)^2 - y^2$$

Notice that there are now two terms, and that each is a perfect square.

Using the rule from the "Factoring the difference of two perfect squares" section earlier in this chapter, $a^2 - b^2 = (a + b)(a - b)$, finish this example:

$$(x + 4)^2 - y^2 = [(x + 4) + y][(x + 4) - y]$$

There's no big advantage to dropping the parenthesis inside the brackets, so leave the answer the way it is.

Knowing When to Quit

One of my favorite scenes from the movie *The Agony and the Ecstasy,* which chronicles Michelangelo's painting of the Sistine Chapel, comes when the Pope enters the Sistine Chapel, looks up at the scaffolding, dripping paint, and Michelangelo perched up near the ceiling, and yells, "When will it be done?" Michelangelo's reply: "When I'm finished!"

The Pope's lament can be applied to factoring problems: "When is it *done*?"

Factoring is *done* when no more parts can be factored. If you refer to the listing of ways to factor two, three, or four (or more) terms, then you can check off the options, discard those that don't fit, and stop when none works. After doing one type of factoring, you should then look at the values in parentheses to see if any of them can be factored.

When factoring, determine what type of expression you have — binomial, trinomial, squares, cubes, and so on. This helps you decide what method to use. Keep going, checking inside all parentheses for more factoring opportunities, until you're done.

✔ Factor $4x^4y - 108xy$.

The GCF of the two terms is $4xy$. Factor that out first.

$$4x^4y - 108xy = 4xy(x^3 - 27)$$

The binomial in the parenthesis is the difference of two perfect cubes and can be factored using the rule from earlier in this chapter.

$$4xy(x^3 - 27) = 4xy(x - 3)(x^2 + 3x + 9)$$

Even though the last factor, the trinomial, seems to be a candidate for unFOIL, you needn't bother when you get the trinomial from factoring cubes because unFOIL can't factor them. The only thing that may factor them is finding a GCF.

You're finished!

✔ Factor $x^4 - 104x^2 + 400$.

There's no GCF, so the only other option when there are three terms is to unFOIL.

$x^4 = x^2 \cdot x^2$ and one pair of factors of 400 is 4 and 100; that's the pair that has a sum of 104.

$$x^4 - 104x^2 + 400 = (x^2 - 4)(x^2 - 100)$$

There are now two factors, but each of them is the difference of perfect squares.

$$(x^2 - 4)(x^2 - 100) = (x + 2)(x - 2)(x + 10)(x - 10)$$

You're finished!

✔ Factor $3x^5 - 18x^3 - 81x$.

The GCF of the terms is $3x$.

$3x^5 - 18x^3 - 81x = 3x(x^4 - 6x^2 - 27)$

The trinomial can be unFOILed.

$3x(x^4 - 6x^2 - 27) = 3x(x^2 - 9)(x^2 + 3)$

The second binomial is the difference of squares.

$$3x(x^2 - 9)(x^2 + 3) = 3x(x - 3)(x + 3)(x^2 + 3)$$

You're finished not only with this problem, but with this chapter!

Part III
Working Equations

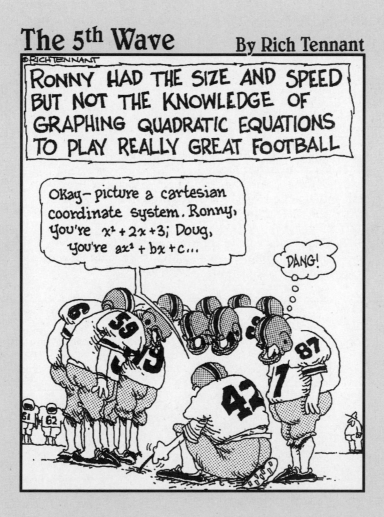

In this part . . .

A re you a fan of Sherlock Holmes or Perry Mason or Jessica Fletcher or the new CSI (Crime Scene Investigators) team? They're all sleuths using investigation and logic and insight to solve problems — and entertain at the same time.

Solving equations can be entertaining, too. The thrill of the hunt and the final correct solution give you that warm, fuzzy feeling all over. On with the show!

Chapter 12

Lining Up for Linear Equations

*I*n this chapter, you find all the different ways to solve linear equations with just two terms. *Linear* means that the highest power of any variable you're solving for is one. Instead of having just an expression with no equals sign (=) and no relationship to anything else, an *equation,* which always has an equal sign, makes a statement about whatever is on one side of the equal sign being equal to the value on the other side. In this chapter, instead of dealing with expressions, such as $3x + 2$, I show you how to solve equations, such as $3x + 2 = 11$.

Two-term equations, unlike two-term presidents, are pretty simple, and you can apply the techniques you use on these equations to more complicated equations.

When you use algebra in the real world, more often than not you turn to a formula to help you work through a problem. Fortunately, when it comes to algebraic formulas, you don't have to reinvent the wheel: You can make use of standard, tried-and-true formulas to solve some common, everyday problems.

Solving with Division

One of the most basic methods for solving equations is to divide each side of the equation by the same number. Many formulas and equations include a coefficient, or multiplier, with the variable. To get rid of the multiplier and solve the equation, you divide. Look at the following example of how to do this:

Solve for the value of x in the equation $20x = 170$.

1. **Determine the multiplier of the variable and divide both sides by it.**

   Because the equation involves multiplying 20 times, undo the multiplication in the equation by doing the opposite of *times*, which is *divide*.

   Divide each side by 20.

   $$\frac{20x}{20} = \frac{170}{20}$$

2. **Reduce both sides of the equal sign.**

   $$\frac{\cancel{20}x}{\cancel{20}} = \frac{170}{20}$$

   $$x = 8.5$$

Do unto one side of the equation what the other side has had done unto it.

Now try a few more examples to get a better understanding.

You need to buy 300 donuts for a big meeting. How many dozen doughnuts is that? Let *d* represent the number of dozen doughnuts you need. There are 12 doughnuts in a dozen, so $12d = 300$. Twelve times the number of doughnuts you need has to equal 300.

1. **Determine the multiplier of the variable and divide both sides by it.**

   Divide each side by 12.

   $$\frac{12d}{12} = \frac{300}{12}$$

2. **Reduce both sides of the equal sign.**

   $$d = 25 \text{ dozen donuts}$$

Archimedes — mover and bather

Born about 287 B.C., Archimedes, an inspired mathematician and inventor, devised a pump to raise water from a lower level to a higher level. These pumps were used for irrigation, in ships and mines, and are still used today in some parts of the world.

He also made astronomical instruments and designed tools for the defense of his city during a war. Known for being able to move great weights with simple levers, cogwheels, and pulleys, Archimedes determined the smallest possible cylinder that could contain a sphere and thus discovered how to calculate the volume of a sphere. The sphere/cylinder diagram was engraved on his tombstone.

A favorite legend has it that as Archimedes lowered himself into a bath, he had a revelation involving how he could determine the purity of a gold object using a similar water-immersion method. He was so excited at the revelation that he jumped out of the tub and ran naked through the streets of the city shouting, "Eureka! Eureka!" (I have found it!)

If your boss makes five times as much money as you do, and her salary is $200,000, what is your salary? (The bigger puzzle may be why she makes so much!)

Write this puzzle as an equation, letting x represent your salary: $5x = 200,000$.

1. **Determine the multiplier of the variable and divide both sides by it.**

   Because the puzzle involves 5 times, undo the multiplication in the puzzle by doing the opposite of *multiplication*, which is *division*.

   Divide each side of the equation by 5.

   $$\frac{5 \cdot x}{5} = \frac{200,000}{5}$$

   $$\frac{\cancel{5} \cdot x}{\cancel{5}} = \frac{200,000}{5}$$

   The 5 in the numerator and the 5 in the denominator cancel each other out on the left. On the right,

   $$\frac{200,000}{5} = 40,000.$$

2. **Reduce both sides of the equal sign.**

   $$x = 40,000$$

The answer to the puzzle is that you make $40,000.

Solving with Multiplication

The opposite operation of multiplication is division. Division was used in the section preceding to solve equations where a number multiplies the variable. The reverse occurs in this section; use multiplication where a number *divides* the variable.

Look at the following example. Try solving for y in $\frac{y}{11} = -2$.

1. **Determine the value that divides the variable and multiply both sides by it.**

   In this case, 11 is dividing the y, so that's what you multiply by.

   $$11\left(\frac{y}{11}\right) = (-2)(11)$$

2. **Reduce both sides of the equal sign.**

   $$\cancel{11}\left(\frac{y}{\cancel{11}}\right) = -22$$

   $$y = -22$$

Try out this puzzle: A wealthy woman's will dictated that her fortune be divided evenly among her nine cats. Each feline got $500,000, so what was her total fortune before it was split up? (Cats don't pay inheritance tax.)

Let f represent the amount of her fortune. Then you can write the equation:

$$\frac{f}{9} = 500,000$$

1. **Determine the value that divides the variable and multiply both sides by it.**

   The fortune divided by 9 gave a share of $500,000.

   In this equation, the fortune was *divided*. Solve the puzzle by *multiplying* each side by 9. The opposite of division is multiplication, so multiplication *undoes* what division did.

   $$9\left(\frac{f}{9}\right) = 500,000 \cdot 9$$

2. **Reduce both sides of the equal sign.**

   $$\cancel{9}\left(\frac{f}{\cancel{9}}\right) = 4,500,000$$
   $$f = \$4,500,000$$

Her fortune was four and a half million dollars. Those are nine very happy kitties.

In the next example, the variable is multiplied by 4 and divided by 5. Solve the problem using multiplication and division.

Solve for a in $\frac{4a}{5} = 12$.

1. **Determine what is dividing the variable.**

   In this case, the 5 is dividing both the 4 and the variable.

2. **Multiply the values on each side of the equal sign.**

   $$5\left(\frac{4a}{5}\right) = 12(5)$$

3. **Reduce and simplify.**

   $$\cancel{5}\left(\frac{4a}{\cancel{5}}\right) = 12(5)$$
   $$4a = 60$$

4. **Now, determine what is multiplying the variable.**

   The number 4 is the coefficient and multiplies the a.

5. **Divide the values on each side of the equal sign.**

   $$\frac{4a}{4} = \frac{60}{4}$$

6. **Reduce and simplify.**

$$\frac{\cancel{4}a}{\cancel{4}} = \frac{60}{4}$$

$$a = 15$$

Solving with Reciprocals

Multiplication and division are opposite operations. Multiplication is *undone* by division and vice versa, as you've seen in the earlier sections. Another option, though, may work better at times — using the *reciprocal* or multiplicative inverse of the number that you're trying to "get rid of." Choose this alternative if a fraction is multiplying the variable, such as in

$$\frac{3x}{19} = 12.$$

Two numbers are *reciprocals* if multiplying them together yields a product of one.

Look at the following examples of reciprocals:

- ✔ 5 and $\frac{1}{5}$ are reciprocals of each other: $5\left(\frac{1}{5}\right) = 1$.
- ✔ $-\frac{3}{7}$ and $-\frac{7}{3}$ are reciprocals: $\left(-\frac{3}{7}\right)\left(-\frac{7}{3}\right) = 1$.
- ✔ The reciprocal of a is $\frac{1}{a}$.
- ✔ The reciprocal of $\frac{1}{b}$ is b.

Solving equations in the fewest possible steps is usually preferable. That's why you can choose to multiply both sides of an equation by $\frac{5}{4}$, the *reciprocal* of $\frac{4}{5}$, to solve for a in the expression $\frac{4a}{5}$, which can be thought of as $\left(\frac{4}{5}\right)a$.

In the following examples, both sides of the equation are multiplied by the reciprocal of the fraction multiplying the variable.

- ✔ In this example, the variable is multiplied by $\frac{4}{5}$.

  $$\frac{4a}{5} = 12$$

  Multiply each side by the reciprocal, $\frac{5}{4}$.

  $$\frac{5}{4}\left(\frac{4a}{5}\right) = \left(\frac{5}{4}\right) \cdot 12$$

  Reduce and simplify.

  $$\frac{5}{\cancel{4}}\left(\frac{\cancel{4}a}{\cancel{5}}\right) = \left(\frac{5}{\cancel{4}}\right) \cdot \cancel{12}^{3}$$

  $$a = 15$$

✔ Solve for x: $\frac{x}{2} = 19$.

$\frac{x}{2}$ is another way of saying $\left(\frac{1}{2}\right) \cdot x$.

So you can solve by multiplying by the reciprocal of $\frac{1}{2}$, which is 2.

$$\frac{\cancel{2}}{1}\left(\frac{1x}{\cancel{2}}\right) = 19\left(\frac{2}{1}\right)$$

$$x = 38$$

✔ Solve for f: $-f = 11$

This is an easy equation to solve, but you may be surprised at how many people get the wrong answer — all because of a little dash in front of a letter. Think of the f as being multiplied by -1. Putting in the -1 gives you a multiplier that you can work with to solve the equation.

What's the reciprocal of -1? It's -1!

$$-f = 11$$

$$(-1)(-1f) = 11(-1)$$

$$f = -11$$

Another example involves using the reciprocal of a fraction, even though it doesn't look like it at first.

Solve for x: $0.7x = 42$. (Remember that 0.7 is the same as .7.)

One way to solve this equation is to divide each side by .7, but sometimes decimal points get misplaced.

A decimal point can get lost easily when it's in the front of a term. You may miss it or think it's a fly speck. Putting a zero in front of a decimal draws attention to the decimal and doesn't change the value of the number. Look at the difference between .8 and writing 0.8 in this sentence.

Solve the equation by dividing:

$$0.7x = 42$$

$$\frac{0.7}{0.7}x = \frac{42}{0.7}$$

$$x = 60$$

Another way to approach this type of problem is to write the decimal as a fraction and multiply by the reciprocal.

To do this in this decimal example, convert 0.7 to $\frac{7}{10}$, thereby replacing the decimal with the fraction. The reciprocal of $\frac{7}{10}$ is $\frac{10}{7}$.

$$\frac{7}{10}x = 42$$

$$\frac{\cancel{10}}{\cancel{7}}\left(\frac{\cancel{7}}{\cancel{10}}\right)x = 42\left(\frac{10}{7}\right)$$

$$x = \frac{\cancel{420}}{\cancel{7}}$$

$$x = 60$$

Setting Up Equations

Solving equations can be fun, especially when the equation has a purpose — to answer a question or solve a problem — and a reality check. This section shows you how to write an equation to answer a question or solve a problem you may have. That's why algebra was developed in the first place — to answer questions and solve problems. Then I discuss whether an answer makes sense or not — doing the reality check.

Changing a written problem into an algebra equation is sort of like translating from one language to another. And the language of algebra has solutions! You need to choose a variable to represent a number in the problem — a number of cats, a number of dimes, and so on. The operations of addition, subtraction, multiplication, and division replace expressions such as *more than, fewer than, times as many,* and so on. For a complete review of how these operations can be used, see Chapter 1.

The following simple sentence can be easily translated into an algebraic statement: *Hugo's nineteen books are three more than twice as many books as Buck has.* The words and expressions to pick up are *nineteen* and *three more than twice.* Let x be the number of books that Buck has. The verb *are* becomes an equal sign (=). *Three more than* is 3 + and *twice* becomes $2x$ because it's twice the number of books. The algebraic equation is $19 = 3 + 2x$.

Now, that wasn't too bad, was it?

Finding a purpose

The following example demonstrates *purpose* (the equation answers a question or solves a problem) and *reality check* (the answer appears to be correct).

A famous rock group called *Aftermath* sold 130,000 copies of their latest CD. This particular CD cost $16. The equation's purpose is to answer the question: What was the total amount of revenue made from the sale of the CDs?

So, what to do? How to solve this?

Letting r represent the total revenue and using the 130,000 and 16, you can set up an equation several different ways. Find the one that's the correct interpretation — the one that solves the problem.

✔ **Choice 1:** Cost of each times total revenue equals number of copies:

$$16 \cdot r = 130,000$$

✔ **Choice 2:** Total revenue divided by the cost of one item equals the number of items sold:

$$\frac{r}{16} = 130,000$$

✔ **Choice 3:** Number of copies divided by total revenue equals cost of each:

$$\frac{130,000}{r} = 16$$

✔ **Choice 4:** Total revenue divided by number of copies equals cost of each:

$$\frac{r}{130,000} = 16$$

You know that multiplying what something sells for times the number of items sold is how you figure out how much money you make, so Choice 2 or Choice 4 works because each requires multiplying the number sold times the cost of each.

In Choice 2, if the total revenue is divided by the cost of each, then the answer is how many CDs were sold. Multiply 16 times 130,000 to get the total revenue.

$$\frac{r}{16} = 130,000$$

$$16 \cdot \frac{r}{16} = 130,000 \cdot 16$$

$$r = 2,080,000$$

In Choice 4, if the number of CDs sold divides the total revenue, then the answer is how much each CD costs. Multiply 130,000 times 16 to get the total revenue.

$$\frac{r}{130,000} = 16$$

$$130,000 \cdot \frac{r}{130,000} = 16 \cdot 130,000$$

$$r = 2,080,000$$

The total revenue is over 2 million dollars.

Doing a reality check

No, don't try to do a reality check here — it's way beyond simple reason to make that much money. The following problem, however, offers a good example of how a reality check can spare you.

The number of soccer players participating at a summer soccer camp is 330 — 11 from each club. You're preparing club participation certificates to give to each club captain, so you need to create an equation to answer the question of how many clubs are represented.

To show you that a reality check can save you from making a big error, pretend that you didn't really think this through and solve the problem with the following equation.

The letter *c* represents the number of soccer clubs.

$$\frac{c}{11} = 330$$

You used the variable and the two numbers in the problem. Does it matter what you use where? Will the equation give you a reasonable answer?

Multiply each side by 11 to solve for *c*.

$$11 \cdot \frac{c}{11} = 330 \cdot 11$$

$$\cancel{11} \cdot \frac{c}{\cancel{11}} = 330 \cdot 11$$

$$c = 3{,}630 \text{ clubs}$$

Humph. This can't be right.

Now, do a reality check. Does the answer make any sense? The answer may satisfy your equation, but if it doesn't make sense, then the equation could be wrong.

Your answer is 3,630 soccer clubs. You're to prepare 3,630 certificates. You realize that something is wrong because only 330 players are involved. You must have made an error.

A quick look at the equation shows that it should have read:

11 players per club times the number of clubs = total number of players

$$11c = 330$$

Now, solve this.

$$11c = 330$$

Divide each side by 11.

$$\frac{11}{11} c = \frac{330}{11}$$

$$\frac{\cancel{11}}{\cancel{11}} c = \frac{330}{11}$$

$c = 30$ clubs

That makes much more sense.

You can solve an equation correctly, but that doesn't mean you chose the right equation to solve in the first place. Make sure that your answer makes sense.

Chapter 13

Solving Linear Equations

· ·

In This Chapter

▶ Simplifying an equation before solving it

▶ Nesting values and operations within grouping symbols

▶ Adding the additive inverse of a number

▶ Knowing when to add, divide, or distribute

▶ Dealing with fractions that are really proportions

· ·

This chapter covers many of the most common methods for solving linear equations, such as adding and subtracting from each side of an equation — keeping things balanced.

Linear describes an equation in which the highest power of any term is one. You don't find powers and roots of variables in linear equations.

Grouping within parenthesis and brackets is covered in this chapter as well as distributing within an equation. And what chapter on solving linear equations would be complete without a section on algebraic fractions? What more could you ask?

Keeping Equations Balanced

Think of an equation as a teeter-totter: Starting out, you have Tweedledum on one side and Tweedledee on the other, and the teeter-totter is balanced perfectly. But, if the Queen of Hearts joins Tweedledum, in order to keep things balanced, you better have the Queen of Diamonds join Tweedledee (as long as she weighs exactly the same as the Queen of Hearts). Likewise, if you decide that the Queens of Diamonds has to go, you need to get rid of the Queen of Hearts, too.

The same principle is true of equations — you keep them balanced, or *true,* by doing exactly the same thing to each side.

An equation is balanced, or *true,* if, when you do the operations indicated, each side comes out to the same answer — even if one side has three terms and the other side of the equal sign has only two terms. Look at the equation $1 + 2 + 3 + 4 = 5 + 5$. Both sides equal ten, even though four terms are on the left and two are on the right.

Whatever else you do to solve an equation, you have to maintain the balance. An equation stays balanced if you add, subtract, multiply, or divide by the same value on each side. Basically, any operation that can be done to numbers can be done to each side of an equation, and it stays balanced.

Playing by the Rules

When you're solving equations with more than three terms, the big question is: "What do I do first?"

Actually, as long as the equation stays balanced, you can perform operations in any order. The following list, however, tells you how to solve your equations in the best order.

The basic process behind solving equations is to use the reverse of the order of operations. The order of operations (see the Cheat Sheet and Chapter 5) is powers or roots first, then multiplication and division, and addition and subtraction last.

Of course, grouping symbols override again. Do the operations necessary to get rid of them first.

So the *reverse* order of operations is

1. **Addition and subtraction:** That means to combine all terms that can be combined both on the same side of the equation and opposite sides.

2. **Multiplication and division:** This is usually the step that isolates or solves for the value of the variable(s).

3. **Multiply exponents and find the roots:** Powers and roots aren't found in these linear equations — they come in quadratic and higher powered equations. But these would come next in the reverse order of operations.

When solving linear equations, the goal is to isolate the variable you're trying to find the value of. Isolating it or getting it all alone on one side can take one step or many steps. And it has to be done according to the rules — you can't just move things willy-nilly without any rules!

Simplifying to keep it simple

ALGEBRA RULES
$$1$$
$$+1$$
$$2$$

To solve linear equations, get all the variables on one side of the equal sign and the numbers on the other side. Then multiply or divide by the coefficient of the variable to solve for it.

✔ **Goal 1:** Get all the variables on one side of the equal sign and numbers on the other side.

To get all the variables on one side and the numbers on the other side, first *simplify*. Simplifying simply means to combine all that can be combined on the same side of the equal sign to put an expression in its most easily understood form. For example,

$$2(3x - 2x + 7) = 30$$

To simplify this problem before solving, simply add, subtract, multiply, and/or divide like terms that are *on the same side of the equal sign*.

So, $3x$ and $2x$ are two like terms on the same side with a minus sign before the $2x$. Simplify by subtracting the $2x$ from the $3x$, which leaves you with x.

$$2(x + 7) = 30$$

Now you can simplify further by distributing the 2 over the x and the 7 *on the same side of the equal sign*.

$$2x + 14 = 30$$

You can't simplify any more but you can subtract 14 from *both* sides of the equation in an effort to get the variable that you're solving for by itself.

$$2x + 14 - 14 = 30 - 14$$

$$2x = 16$$

✔ **Goal 2:** Multiply or divide by the multiplier of the variable to solve for it or get it alone.

To satisfy Goal 2, divide both sides of the equation by 2.

$$\frac{2x}{2} = \frac{16}{2}$$

$$x = 8$$

Check your answer by substituting 8 for x in the original equation to see whether it's a real solution.

$$2[3(8) - 2(8) + 7] = 30$$

$$2[24 - 16 + 7] = 30$$

$$2[15] = 30$$

$$30 = 30$$

Nesting isn't for the birds

When you have a number or variable that needs to be multiplied by every value inside parentheses, brackets, braces, or a combination of those grouping symbols — you *distribute* that number or variable. Distributing means that the number or variable next to the grouping symbol multiplies every value inside the grouping symbol. (Distribution is covered in Chapter 8 and discussed later in this chapter.) If two or more of the grouping symbols are inside one another, they're *nested*. Following are some examples of nested expressions.

$$\{3x - [4 + (2x - 1)] \cdot 5\}$$

$$(4xy\{3 + 5x[2(x^2 + 11) - 5] + (x^2 + 1)\} - 2)$$

Nested expressions are written within parentheses, brackets, and braces to make the intent clearer. The following conventions are used when nesting:

1. When using nested expressions, every *opening* grouping symbol, such as left parenthesis (, bracket [, or brace { has to have a *closing* grouping symbol — a right parenthesis), bracket], or brace }.

2. When simplifying nested expressions, work from the inside to the outside. The innermost expression is the one with no grouping symbols inside it. Simplify that expression or distribute over it so the innermost grouping symbols can be dropped. Then go to the next innermost grouping.

Work from inside grouping symbols to the outside. When grouping symbols are not used, refer to the order of operations (see Chapter 5 and the Cheat Sheet in the front of this book).

Solve for x in the following equations to get a clear picture of what working with grouping symbols is like.

1. **Look for a grouped expression with no grouping symbols inside and distribute.**

   $$6[2 - 3(x - 4)] = 5[2(x + 1) - 2]$$

   The $(x - 4)$ and $(x + 1)$ both fit this description.

   Each of the expressions in parentheses has a number times it, so distribute that number.

   $$6[2 - 3x + 12] = 5[2x + 2 - 2]$$

2. **Simplify any terms that result from distributing.**

   Simplify inside the brackets by doing the operations indicated.

   $$6[14 - 3x] = 5[2x]$$

3. **Look for other grouped expressions with no grouping symbols inside, distribute, and simplify.**

   The two brackets each have a number multiplying them, so distribute again.

   $$84 - 18x = 10x$$

   Add the same term to each side.

   $$84 - 18x + 18x = 10x + 18x$$
   $$84 = 28x$$

4. **Solve the equation.**

   Divide each side by the same number.

   $$\frac{84}{28} = \frac{28x}{28}$$
   $$3 = x$$

Try to work through the following example of grouping operations on your own. Then check your work against the steps that follow to give yourself a little more practice.

1. Look for a grouped expression with no grouping symbols inside and distribute.

   $$5\{[3(5 - x) + 2] - 10\} = 6x + 7[2(x - 1) + 12]$$

   A grouped expression is inside each bracket on each side of the equation. Distribute over the parentheses inside the brackets because numbers multiply them.

   $$5\{[15 - 3x + 2] - 10\} = 6x + 7[2x - 2 + 12]$$

2. Simplify any terms that result from distributing.

   $$5\{[17 - 3x] - 10\} = 6x + 7[2x + 10]$$

3. Look for other grouped expressions with no grouping symbols inside, distribute, and simplify.

   On the left, the brackets can be dropped because there's nothing multiplying them. On the right side, distribute the 7.

   $$5\{17 - 3x - 10\} = 6x + 14x + 70$$
   $$5\{7 - 3x\} = 20x + 70$$
   $$35 - 15x = 20x + 70$$

4. Solve the equation.

   Add $15x$ to each side.

   $$35 - 15x + 15x = 20x + 15x + 70$$
   $$35 = 35x + 70$$

Add −70 to each side.

$$35 - 70 = 35x + 70 - 70$$

$$-35 = 35x$$

Divide each side by 35.

$$\frac{-35}{35} = \frac{35x}{35}$$

$$-1 = x$$

Balancing by addition

Use addition to solve an equation when one or more terms are on one side of the equal sign along with a variable. Using addition or subtraction makes sense because addition or subtraction separates one term from another.

For example, the equation $x - 3 = 8$ has three terms, two on the left and one on the right. You could choose to do many different things to each side, but the *best* choice is to add 3 — the additive inverse of −3 — to each side.

Adding 3 is the best choice because a number and its additive inverse add up to zero, which leaves nothing except the variable on one side of the equation, and that's something to aim for — getting the variable by itself on one side of the equal sign.

$$x - 3 = 8$$

$$x - 3 + 3 = 8 + 3$$

$$x = 11$$

You still have the same truth or relationship because it's balanced. And now the answer is staring you right in the face.

The following example shows why and how to use additive inverses when solving equations.

✔ Suppose that you collect state quarters. This evening you have 7 more quarters than you had when you left this morning. If you have 11 quarters now, how many did you have when you left this morning?

Letting x represent the unknown number of quarters you had this morning (remember that variables always represent numbers) is a good choice because that's the number you want to solve for. After solving the equation, you'll have the answer right there.

Quarters this morning + 7 quarters = 11 quarters now

$$x + 7 = 11$$

Adding the inverse of + 7 to each side is the same as subtracting 7 from each side.

$$x + 7 - 7 = 11 - 7$$

$$x = 4$$

Your answer is $x = 4$. The variable x represents the 4 quarters you had this morning.

✔ Work through this example and solve for x to give you an idea of how you can get a negative answer.

$$x + 11 = 4$$

$$x + 11 - 11 = 4 - 11$$

$$x = -7$$

Yes, you *can* have a *negative* answer. This doesn't work on teeter-totters, but it works in algebra.

Adding first versus dividing

As with many things in life, there are easier, harder, and downright ugly ways of doing these equations. Sometimes avoiding *ugly* is not possible, but as long as you do the same thing to each side of an equation it stays balanced. Also, you don't have to use the methods in any particular order — some orders are just recommended.

✔ Say, that just this once, you decide to use the division method before addition in the equation $4x + 51 = 83$.

Divide each side by four.

$$\frac{4x}{4} + \frac{51}{4} = \frac{83}{4}$$

Then you have to add fractions to each side and simplify the answer. This isn't too bad to work with, but the fractions are unnecessary. Creating fractions before you need to is something you should avoid.

Distributing first versus dividing

Distributing and multiplying first can be much easier than dividing, and the following example shows you how.

I was in a fishing contest (yes, there *are* those kinds of things, and some people take them very seriously!), which is how I came up with this example:

Four of my friends each got 8 more fish than I did. (Better bait?) But that didn't compare to the winner, some guy from up north, who got 6 times as many as I did. If you take away 2 of the winner's fish, that still equaled the *total* that my four friends caught altogether. So how many fish did I catch? How many did my friends get?

Let x represent the number of fish that I caught. (Size doesn't matter, thank goodness.)

$x + 8$: each friend got 8 more than I did

$4(x + 8)$: 4 friends did this

$6x$: the winner got 6 times as many as I did

$6x - 2$: take 2 away from the winner's amount

Four friends = Winner − 2

$4(x + 8) = 6x - 2$

There are two ways to approach this problem:

1. You could divide each side by 4 to get rid of the multiplication.

2. You could *distribute* the 4 and avoid fractions.

I'll show you the steps to work through both approaches, and you can decide which makes more sense in this case.

1. **Divide each side by 4.**

   This is sort of nasty, making fractions already.

   $$4(x + 8) = 6x - 2$$

2. **Divide each term by 4.**

   $$\frac{4(x + 8)}{4} = \frac{6x}{4} - \frac{2}{4}$$

3. **Simplify.**

   $$x + 8 = \frac{6x}{4} - \frac{2}{4} = \frac{3}{2}x - \frac{1}{2}$$

4. **Add −8 to each side.**

   $$x + 8 - 8 = \frac{3}{2}x - \frac{1}{2} - 8$$

   $$x = \frac{3}{2}x - 8\frac{1}{2}$$

5. Add $-\dfrac{3}{2}x$ to each side.

$$x - \frac{3}{2}x = \frac{3}{2}x - \frac{3}{2}x - 8\frac{1}{2}$$

$$x - \frac{3}{2}x = -8\frac{1}{2}$$

6. Simplify.

$$-\frac{1}{2}x = -8\frac{1}{2}$$

7. Multiply each side by –2.

$$(-2)\left(-\frac{1}{2}x\right) = \left(-8\frac{1}{2}\right)(-2)$$

REMEMBER

When multiplying mixed numbers, you need to change them to improper fractions, first. For example, $8\frac{1}{2}$ becomes $\frac{17}{2}$. (See Chapter 3 for more on fractions.)

$$(-2)\left(-\frac{1}{2}x\right) = \left(-\frac{17}{2}\right)(-2)$$

$$x = 17$$

I caught 17 fish.

Each of my friends caught 17 + 8, or 25, fish.

The winner caught 6 × 17, or 102 fish.

Okay, so the first method took seven steps to get to the answer and made you work with fractions right off the bat. Choosing the distribution method makes the problem quite a bit more civilized. Remember, you're starting with $4(x + 8) = 6x - 2$.

1. Distribute the 4 over the two terms.

$$4x + 32 = 6x - 2$$

2. Add $-4x$ to each side.

$$4x - 4x + 32 = 6x - 4x - 2$$

$$32 = 2x - 2$$

3. Add 2 to each side.

$$32 + 2 = 2x - 2 + 2$$

$$34 = 2x$$

4. Divide both sides by 2.

$$\frac{34}{2} = \frac{2x}{2}$$

$$17 = x$$

Four little steps and you're there! This is a much nicer method. I don't know about you, but I surely do prefer not having to deal with fractions when it isn't necessary.

When distributing, be careful with negative signs. Losing or misplacing a negative sign is easy to do, so write and use them carefully.

✔ Solve for x in $5 - 2(x - 3) = 4x - 7$

Distribute the -2 through on the left.

$$5 - 2x + 6 = 4x - 7$$

Simplify the terms on the left.

$$-2x + 11 = 4x - 7$$

Add $-4x$ to each side and -11 to each side.

$$-2x - 4x + 11 - 11 = 4x - 4x - 7 - 11$$

$$-6x = -18$$

$$-\frac{6x}{6} = -\frac{18}{6}$$

$$x = 3$$

Now check it.

$$5 - 2(x - 3) = 4x - 7$$

If $x = 3$, then

$$5 - 2(3 - 3) = 4(3) - 7$$

$$5 - 0 = 12 - 7$$

Looks good to me!

Balancing Fractions

Fractions pop up all over the place. Some people cringe at the thought of them; others just take them in stride. Still other people really *like* them. This section gives you some practical ways of dealing with fractions because fractions are a fact of life — and linear equations.

Multiplying cross products

A *proportion* is two ratios that are equal. A proportion can be written as an equation with colons, $3{:}4 = 6{:}8$ or as an equation with fractions, $\frac{3}{4} = \frac{6}{8}$. Comparing sizes and amounts is a common situation in algebra and in life. If four men for every nine women are in the room, then all sorts of social problems can come to mind.

A *proportion* is the relationship between one pair of numbers, *a* and *b*, and another pair of numbers, *c* and *d*, such that you get the same results when they are divided (*a* divided by *b* and *c* divided by *d* — $\frac{a}{b} = \frac{b}{a}$). The result of multiplying the numerator of one fraction times the denominator of another fraction is a *cross product*. If the two cross products of two fractions are equal ($a \cdot d = b \cdot c$), then you have a *proportion*, which is the same as equivalent fractions.

To demonstrate this, look at these cross products of some equivalent fractions.

$$\frac{1}{2} = \frac{3}{6} \qquad = \qquad 1 \cdot 6 = 3 \cdot 2 \qquad = \qquad 6 = 6$$

$$\frac{3}{4} = \frac{9}{12} \qquad = \qquad 3 \cdot 12 = 9 \cdot 4 \qquad = \qquad 36 = 36$$

$$\frac{8}{10} = \frac{12}{15} \qquad = \qquad 8 \cdot 15 = 12 \cdot 10 \qquad = \qquad 120 = 120$$

Notice that only one term is on each side of the equal sign when you finish. A proportion must not have more than one term on each side of the equals sign. Look at the following problems to see how you can use proportions to solve equations.

✔ Solve for x if $\frac{x}{8} = \frac{5}{20}$

$$x \cdot 20 = 8 \cdot 5 = 40$$

Divide each side by 20

$$\frac{x \cdot \cancel{20}}{\cancel{20}} = \frac{40}{20}$$

$$x = 2$$

The proportion is

$$\frac{2}{8} = \frac{5}{20}$$

✔ The chocolate chip cookie recipe calls for 2¼ cups of flour and 2 eggs. You want to use all 12 eggs in your refrigerator. How much flour will you need (assuming you have enough of all the other ingredients, too)?

The proportion is $\dfrac{2\frac{1}{4}\,cups\,flour}{2\,eggs} = \dfrac{x\,cups\,flour}{12\,eggs}$

There are several ways to write this proportion or one equivalent to it. For example, the answer can also be obtained with:

$$\frac{2\frac{1}{4}\,cups\,flour}{x\,cups\,flour} = \frac{2\,eggs}{12\,eggs}$$

$$\frac{12\,eggs}{2\frac{1}{4}\,cups\,flour} = \frac{x\,cups\,flour}{2\,eggs}$$

In each case, the cross products are the results of the same values being multiplied together.

Back to the cookie recipe problem. The proportion is

$$\frac{2\frac{1}{4}}{2} = \frac{x}{12}$$

$$2\frac{1}{4} \cdot 12 = 2 \cdot x$$

$$\frac{9}{4} \cdot 12 = 2x$$

The mixed number was changed to an improper fraction.

$$27 = 2x$$

$$x = \frac{27}{2} = 13\frac{1}{2} \text{ cups of flour}$$

That's a lot of cookies!

Transforming fractions into proportions

Proportions are fairly easy to deal with. Multiplying numerators and denominators to find cross products makes them a piece of cake. Other types of equations with fractions can give us more of a challenge, but nothing that can't be handled!

$$\frac{x}{3} = \frac{8}{45} \text{ is a proportion.}$$

$$\frac{x}{3} + \frac{4}{5} = \frac{8}{45} \text{ is } not \text{ a proportion.}$$

But you can *make* a proportion. After all, proportions are so easy to work with — it's a familiar process.

✔ To make the following equation into a proportion, you have to change the problem so just one fraction is on the left, not two. If you need to review adding fractions, see Chapter 3.

$$\frac{x}{3} + \frac{4}{5} = \frac{8}{45}$$

To add the fractions on the left together, find the common denominator for the fractions, which is 15 in this case.

Create equivalent fractions with a denominator of 15.

$$\frac{5 \cdot x}{5 \cdot 3} + \frac{3 \cdot 4}{3 \cdot 5} = \frac{8}{45}$$

$$\frac{5x}{15} + \frac{12}{15} = \frac{8}{45}$$

Add the fractions together.

$$\frac{(5x + 12)}{15} = \frac{8}{45}$$

Now you have a proportion and can multiply to find the cross products and get rid of the fraction.

$$(5x + 12) \cdot 45 = 15 \cdot 8$$

Distribute the 45 and simplify the right side. You may prefer dividing each side by 15 first to keep the numbers smaller. That's up to you.

$$225x + 540 = 120$$

Subtract 540 from each side.

$$225x + 540 - 540 = 120 - 540$$

$$225x = -420$$

Divide each side by 225 and reduce the fraction on the right.

$$\frac{225x}{225} = \frac{-420}{225}$$

$$x = \frac{-28}{15}$$

At the beginning of this problem, it may have been easier to move the term without the variable to the other side first, but that's a matter of personal choice. The result is the same.

Keeping fractions

Fractions may not be your very favorite things in the world, but don't go out of your way to avoid them. Sometimes their presence is unavoidable. The following section offers you some reasonable *game plans* to make them workable and to give you the best chance at the correct answer. The two best strategies are

1. Multiply each side of the equal sign by a number that gets rid of the denominators of the fractions.

2. Combine all the fractions on each side separately to create a proportion.

The following situation is good for starters. Use fractions to solve the problem.

If Anki can clean the house in 6 hours, and Bernardo can clean the house in 12 hours, but it takes Clara 15 hours to clean the house, then how long will it take if all three pitch in and clean the house together (with no bickering or slacking off)?

The equation and fractions I would use are

$$\frac{x}{6} + \frac{x}{12} + \frac{x}{15} = 1$$

If x is how many hours it'll take to do the house working together then $\frac{x}{6}$ represents $\frac{1}{6}$ of the job in one hour (Anki's rate), $\frac{x}{12}$ represents $\frac{1}{12}$ of the job in one hour (Bernardo's rate), $\frac{x}{15}$ represents $\frac{1}{15}$ of the job in one hour (Clara's rate), and the sum of all the "parts" they do is what is accomplished in an hour.

Anki's rate = $\frac{x}{6}$

Bernardo's rate = $\frac{x}{12}$

Clara's rate = $\frac{x}{15}$

To solve this, you could go the proportion route, and what is actually done here will resemble it, but another plan of attack is to find a common denominator for the three fractions on the left *and* for the number on the right. Then you can multiply each side of the equation by that common denominator and get rid of the fractions completely.

The common denominator for fractions with denominators of 6, 12, and 15 is 60. How did I get this? One way is to guess. Another way is to write the prime factorizations of the numbers and find what they're all common to. Another quick trick is given in the following steps.

1. **Find the least common denominator by taking the biggest of the denominators and checking all its multiples until you find one that all the denominators divide.**

    In the case of this problem, 15 is the biggest denominator:

    $15 \times 1 = 15$: Neither 6 nor 12 divide that evenly.

    $15 \times 2 = 30$: Only the 6 divides that evenly.

    $15 \times 3 = 45$: Neither 6 nor 12 divide that evenly.

    $15 \times 4 = 60$: A winner!

2. **Rewrite the problem with the common denominator:**

    $$\frac{x}{6} + \frac{x}{12} + \frac{x}{15} = 1$$

    $$\frac{10 \cdot x}{10 \cdot 6} + \frac{5 \cdot x}{5 \cdot 12} + \frac{4 \cdot x}{4 \cdot 15} = 1$$

    $$\frac{10x}{60} + \frac{5x}{60} + \frac{4x}{60} = 1$$

3. **Multiply each side by the common denominator.**

    When you multiply by 60 in the sample problem, all the denominators divide out or disappear.

    $$10x + 5x + 4x = 60$$

    Then you can add the terms on the left side:

    $$19x = 60$$

    $$\frac{19x}{19} = \frac{60}{19}$$

    You end up with $x = \frac{60}{19}$ or $3\frac{3}{19}$ hours, which is about 3 hours and 9 minutes.

See what can be accomplished when people work together!

If fractions are on each side of the equal sign, you have two choices:

1. Find the common denominator for each side, combine the fractions, and solve the proportion that results.

2. Find a number to multiply both sides by to get rid of the fractions. The choice here is going to depend on how nice the fractions are.

If you have a problem, such as

$$\frac{x}{13} + \frac{2x}{47} = \frac{5}{9} + \frac{x}{37}$$

then it would keep the numbers smaller to just find a common denominator for each side separately and do the proportion. The numbers will be awful, no matter what you decide to do.

An equation that lends itself more to finding one, single common denominator for the whole problem is shown here, in the next example.

 $\frac{3x}{4} + \frac{2x}{5} = \frac{17x}{30} + \frac{x}{2} + 1$

All the fractions have denominators that divide 60, so that will be the common denominator for both sides.

$$\frac{45x}{60} + \frac{24x}{60} = \frac{34x}{60} + \frac{30x}{60} + \frac{60}{60}$$

Multiply each side of the equation by 60. The 60s in the denominators all divide out leaving:

$$45x + 24x = 34x + 30x + 60$$

$$69x = 64x + 60$$

Subtract 64x from each side.

$$69x - 64x = 64x - 64x + 60$$

$$5x = 60$$

Divide each side by 5.

$$x = 12$$

When reducing fractions, *every* term has to contain the factor being divided out.

Reducing fractions is fine, as long as the reducing is done correctly. Dividing the numerator and the denominator by the same number can make a fraction simpler. But, if the numerator or denominator have more than one term, then *each term* has to be divided by that number.

Solving for Variables

Sometimes algebraic answers are variables. Sometimes it's a good idea to leave the answer *open* or flexible; it's left as a variable. Variables are useful when a situation needs to be used over and over to solve a problem. The "variable answer" does become a number when the other variables in the problem are given values. Look at this example.

✔ You got an 80 on the first test and a 40 on the second test. You have another test tomorrow! You have to average 90 on your three tests to get permission to go skiing. You have to average 80 to keep from losing driving privileges. You have to average 70 to keep from losing phone privileges. You have to average . . . You get the picture.

Add up the scores and divide by the number of scores to find an *average*. So the

$$\text{Average of 3 tests} = \frac{(\text{Score on test 1} + \text{Score on test 2} + \text{Score on test 3})}{3} = \frac{(80 + 40 + \text{Score on test 3})}{3}$$

Let the average of the three tests be V (V is for "victory," of course!). Let x represent the score on that third test.

$$V = \frac{(80 + 40 + x)}{3}$$
$$V = \frac{(120 + x)}{3}$$

You'll be getting a variable in your answer because there are two variables in the problem that you begin with. As long as more than one variable is in the original equation, a variable will be in the answer; your answer won't have a simple, single number.

Because you want to know what *score* you need on the next test, solve for x.

$$V = \frac{(120 + x)}{3}$$

Multiply each side by 3.

$$3 \cdot V = \frac{3 \cdot (120 + x)}{3}$$
$$3V = 120 + x$$

Subtract 120 from each side.

$$x = 3V - 120$$

This answer has a variable in it. Now apply it.

If x is the score you need to get an average of V, and if $x = 3V - 120$, then you can decide on your target average, V, and determine the test score needed. Try this out and find out what you get to do. You want an average of 90 so you can go skiing.

$$x = 3V - 120$$

Replace the V with 90.

$$x = 3(90) - 120$$

$$x = 270 - 120 = 150$$

Oops! The test is only worth 100 points. Can't get more than it's worth. Scrap that idea. No skiing this time.

How about an average of 80 so you can drive to the mall?

$$x = 3V - 120$$

Replace the V with 80.

$$x = 3(80) - 120$$

$$x = 240 - 120 = 120$$

Guess that won't work either. Saved some money, anyway.

How about an average of 70? Without the phone you'll wither and die!

$$x = 3V - 120$$

Replace the V with 70.

$$x = 3(70) - 120$$

$$x = 210 - 120 = 90$$

Finally, a score that's possible. At least, it's possible on paper. Guess you shouldn't have gotten in that jam in the first place!

Now that you made it through one test, try a problem that presents more of a challenge.

✔ Solve for x in $4x - tx = 16$

Factor out the x on the left.

$$x(4 - t) = 16$$

Divide each side by $4 - t$

$$\frac{x(4 - t)}{(4 - t)} = \frac{16}{(4 - t)}$$

$$x = \frac{16}{(4 - t)}$$

✔ Solve for x in $5(3A + x) = 20A - 1$

You have two ways to proceed, but try distributing the 5 on the left. If you divide by 5, you end up with fractions right away. Save that until really necessary.

$$5(3A + x) = 20A - 1$$

$$15A + 5x = 20A - 1$$

Subtract $15A$ from each side.

$$15A - 15A + 5x = 20A - 15A - 1$$

$$5x = 5A - 1$$

Divide each side by 5.

$$\frac{5x}{5} = \frac{(5A - 1)}{5}$$

$$x = \frac{(5A - 1)}{5}$$

This is your answer, but look at these equations:

$$\frac{(5A - 1)}{5} \neq \frac{(\cancel{5}A - 1)}{\cancel{5}}$$

$$\frac{(5A - 1)}{5} \neq A - 1$$

If the numerator or denominator has more than one term, then *each* term has to be multiplied by what is being divided out.

The fraction preceding *doesn't* reduce because a factor of 5 in –1 doesn't exist. The following fraction, however, *does* reduce:

$$\frac{(5A - 10)}{5} = \frac{5(A - 2)}{5}$$

$$\frac{\cancel{5}(A - 2)}{\cancel{5}} = A - 2$$

Getting Impossible Answers

There are nice numbers for answers, and there are *impossible* answers. These answers occur most often when doing problems that have fractions and radicals in them. Fractions can cause problems because you aren't allowed to put a zero in the denominator. You'd never do that, you say. Well, of course not, but what if you didn't mean to? What if a variable is in the denominator and you accidentally let the variable equal zero? Here are some ways to avoid such an "accident."

Being careful when working with fractions is especially important because sometimes you can do everything right but get an impossible answer. You want to be careful not to let the denominator equal zero.

✔ You proclaim that $x = 1$. You're saying that x is a variable which represents the number 1 right now, at this time.

   Square both sides of the equation

   $$x^2 = 1$$

   Now subtract 1 from each side

   $$x^2 - 1 = 1 - 1$$
   $$x^2 - 1 = 0$$

   Divide each side by $x - 1$

   $$\frac{(x^2 - 1)}{(x - 1)} = \frac{0}{x - 1}$$

   $$\frac{(x^2 - 1)}{(x - 1)} = 0$$

   The numerator on the left side is the difference of perfect squares, so it factors into $(x - 1)(x + 1)$.

   $$\frac{(x - 1)(x + 1)}{(x - 1)} = 0$$

   Reduce the fraction on the left.

   $$\frac{\cancel{(x - 1)}(x + 1)}{\cancel{(x - 1)}} = 0$$

   Leaving

   $$x + 1 = 0.$$

   If you subtract 1 from each side, $x + 1 - 1 = 0 - 1$

   $$x = -1$$

   But you proclaimed at the beginning that $x = 1$! x can't be both $+1$ and -1. A number can't be two different things at the same time. So what happened?

What you did was an *illegal maneuver*. You shouldn't have divided by $x - 1$ because that's equal to 0 if $x = 1$. Sometimes you can see that you're dividing by 0, such as in this case. Sometimes it's not as obvious. That's why you have to be sure to check answers because you can get an impossible or nonsense answer. Sometimes it's obvious that the answer is impossible. Sometimes it's not.

The following examples illustrate what can happen. This time, don't do anything *illegal*.

- ✔ Solve for x in $\frac{(x+1)}{x} = 1$

  Cross-multiply to get

  $$x + 1 = x$$

  Subtract x from each side.

  $$1 = 0$$

  That's just not possible, so this problem doesn't have a solution.

- ✔ Solve for x in $\frac{(9x+4)}{x(x+2)} = \frac{7}{(x+2)}$

  Change the right side so it has the same denominator as that on the left side. You can do that by multiplying the numerator and denominator by x.

  $$\frac{(9x+4)}{x(x+2)} = \frac{7x}{x(x+2)}$$

  Now multiply each side by that common denominator and you have only the numerators (tops) left.

  $$9x + 4 = 7x$$
  $$9x - 9x + 4 = 7x - 9x$$
  $$4 = -2x$$
  $$\frac{4}{-2} = \frac{-2x}{-2}$$
  $$x = -2$$

  Now this looks like a perfectly respectable answer, doesn't it? The only problem comes in when you go to check it. Look at what happens when you replace each x with -2.

  $$\frac{(9(-2)+4)}{-2(-2+2)} = \frac{7}{-2+2}$$
  $$-\frac{14}{0} = \frac{7}{0}$$

  You ended up with zeros in the denominators. You can't divide *anything* by 0. Sorry. No answer. This is impossible! Again, there's no solution.

If you keep the equation balanced and follow the basic rules for solving linear equations, you won't go wrong. Check your answer to be sure that you end up where you want to be!

Chapter 14

Taking a Crack at Quadratic Equations

Quadratic (second-degree) equations are nice to work with because they are manageable. Finding the solution or deciding whether a solution exists is relatively easy. A *quadratic equation* is a quadratic expression with an equal sign attached. It's also called a *second-degree* equation because the highest power in the equation is two — the degree is the second power. As with linear equations, methods, given in detail in this chapter, are used to solve quadratic equations. The main technique for solving these equations is by factoring, but there's also a quick and dirty rule for one of the special types of quadratic equations. And, just because someone puts in some numbers and makes up a quadratic equation, that doesn't mean there's a solution or answer to it. You'll be able to tell if there's no answer because the methods will show you.

Quadratic equations are basic to algebra and many other sciences. Some equations say that what goes up must come down. Other equations describe the paths that planets and comets take. In all, quadratic equations are fascinating — and almost as much fun to work with.

Squaring Up to Quadratics

A *quadratic equation* is one containing a term with an exponent of two and no term with a higher power.

A quadratic equation has a general form that goes like this:

$$ax^2 + bx + c = 0$$

If this looks familiar, it means that you've read Chapter 10, which talks about factoring and working with quadratic expressions, or that you see visions of quadratic equations dance through your head. Remember, an expression is comprised of one or more terms, but has no equals sign. Adding an equals sign changes the whole picture: Now you have an equation that says something. It says something that's a true statement if the solutions are put in for the variables. It says something that's false if the wrong numbers are put in for the variables.

The a in the general quadratic equation can be any real number except zero. If a were zero, there wouldn't be a term with a power of two, and it wouldn't be quadratic any more. The b and c can also be any number, and they *can* be equal to zero. You can have one or both of those terms missing and still have a quadratic equation. Look at some examples:

- $4x^2 + 3x - 2 = 0$: In this equation, none of the coefficients are zero.
- $2x^2 - 5 = 0$: In this equation, the b is equal to zero.
- $x^2 + 3x = 0$: In this equation, the c is equal to zero.
- $x^2 = 0$: In this equation, both b and c are equal to zero.

A special feature of quadratic equations is that they can, and often do, have two answers. Yes, you can have two completely different answers to just one little problem.

- The following quadratic equation has two answers:

    $$x^2 - 5x = 6$$

    This is quadratic because of the x^2 term. The two answers that work in this are 6 and −1.

    $$(6)^2 - 5(6) = 36 - 30 = 6, \text{ and } (-1)^2 - 5(-1) = 1 + 5 = 6$$

    Both of them work!

How did I do that? I used the methods on solving quadratic equations that are found later in this chapter. My goal in this section is to get you used to having two different answers that work. You can jump ahead if your curiosity is getting the best of you.

How can an equation have two answers? Which is *right* in an application or story problem? For instance, if the story problem asks about how much something costs, how can there be *two* correct answers? Well, sometimes there *are* two right answers to the quadratic equation, but usually one of the answers doesn't really make sense in the particular application. The nonsensical answer does solve the equation and just comes along as extra baggage. You just have to make a decision as to whether to pay attention to the extra answer.

✔ Look at this equation that has two answers:

The following equation is used to find out how high a ball is t seconds (t stands for time in seconds) after it's thrown upward from a 16-foot-high wall (h stands for the height of the ball).

$$h = -16t^2 + 80t + 16$$

Don't worry about where I got the equation; it's something discussed in physics and in many math classes.

I want to figure out when the ball is 80 feet above the ground, so I put in 80 for the height, h.

$$80 = -16t^2 + 80t + 16$$

I just happen to know that when t is equal to 1 or 4, the equation is true. Well, I don't actually know that, but I used the methods from this chapter for finding the solution of a quadratic equation to get the answers. Again, I'm showing you how two answers can work and both make sense.

When $t = 1$,

$$80 = -16(1)^2 + 80(1) + 16 = -16 + 80 + 16 = 80,$$

and when $t = 4$

$$80 = -16(4)^2 + 80(4) + 16 = -256 + 320 + 16 = 80$$

Both work! So this equation says that when t equals 1 *and* when t equals 4 the ball is 80 feet in the air.

If you throw a ball up into the air from 16 feet high, then the ball could go up, pass the 80 foot level, go higher than that, and then be at the 80 foot level again on the way down.

The following quadratic equation has two answers, but only one makes any sense in the actual problem. The answers both *work* in the equation, but only one answers the question.

✔ The quadratic equation $C = 0.04n^2 + 2n + 100$ tells you how much it costs (C) to produce a number (n) of whachamacallits.

$C = 0.04n^2 + 2n + 100$ gives you the total cost C for n of them, and you want to know when the cost is going to be $124.

Replace the C with 124 and get $124 = 0.04n^2 + 2n + 100$.

If $n = 10$ or if $n = -60$, either will make the equation into a true statement. Getting these two answers — one of them sometimes negative — happens frequently when you use an equation to model what happens in real life. The equation usually works wonderfully to give you answers, but you can't use it beyond what's reasonable. With this particular equation, it wouldn't make sense to use negative numbers for n. And it also may not be reasonable to use values of n up in the billions or trillions. The price that's paid for using these nice equations is that they have to be used under reasonable circumstances.

When $n = 10$,

$$0.04(10)^2 + 2(10) + 100 = 0.04(100) + 20 + 100 = 4 + 120 = 124$$

When $n = -60$,

$$0.04(-60)^2 + 2(-60) + 100 = 0.04(3600) - 120 + 100 = 144 - 20 = 124$$

So both work! But that doesn't make any sense. You can't produce a negative number of items. Often, quadratic equations are very nice if you just use half or part of them — the part that makes sense in a certain situation.

Rooting Out Another Result from Quadratic Equations

The general quadratic equation has the form $ax^2 + bx + c = 0$, and b or c or both can be equal to zero. This section shows you how nice it is when b is equal to 0. The solutions are fairly easy to find.

The first 20 perfect squares (products of a number times itself) are: 1, 4, 9, 16, 25, 36, 49, 64, 81, 100, 121, 144, 169, 196, 225, 256, 289, 324, 361, 400.

Notice that the numbers go from a low of one to a high of 400. There aren't any other perfect squares between the ones listed. That means that the other 380 numbers between one and 400 are *not* perfect squares. The perfect squares all have nice square roots. The square root of 121 is 11; the square root of 256 is 16. Isn't that nice? But the square root of 200 isn't nice at all; it's an *irrational* number. Irrational numbers do not terminate or repeat themselves after the decimal point. For example, the square root of two, an irrational number, is 1.414213562373 . . . An irrational number can't ever be written as a fraction. They're just as their name describes: wild and irrational. The roots have decimal values that can be approximated with a calculator, if you need.

You can find a discussion of squares and square roots in Chapter 4, and a table of squares and cubes on the Cheat Sheet at the front of the book.

Don't worry if you don't recognize some of the larger squares because they aren't used frequently, and you usually get some sort of a hint that the number is a perfect square when you're doing a problem. Sometimes the hint comes from the wording of the problem – it may talk about a square room or sides of a right triangle. Sometimes the hint is just that it'd be so nice if it was square. (You can check the Cheat Sheet for a list of squares.)

And here's a twist to square roots and squares. Usually, if you're asked for the square root of 25, you say, "Five." Well, that's right of course, but that's just the *principal square root,* which isn't the only square root. Did you forget something? No, you just have to think outside the box. What other number times itself gives you 25? Why, –5, of course. So, when solving equations, the two roots of 25 are +5 and –5.

The *principal square root* of a number is just the positive number that, when multiplied by itself, gives you the original number. The principal square root of 25 is 5. When doing a square root to solve an equation, both the principal square root and its inverse (the negative one) are used.

So, under certain circumstances, such as solving quadratic equations, you have to consider that *other* answer, also.

The following is the rule for some special quadratic equations — the ones where $b = 0$. They start out looking like $ax^2 + c = 0$, but the c is usually negative and its inverse gets added to each side of the equation so the equation looks like $ax^2 = c$.

Square Root Rule: If $x^2 = k$, then $x = \pm \sqrt{k}$.

If the square of a variable is equal to the number k, then the variable is equal to the principal square root of k or its opposite.

The following examples show you how to use this square root rule on quadratic equations where $b = 0$.

- ✔ Solve for x in $x^2 = 49$

  Using the square root rule, $x = \pm \sqrt{49} = \pm 7$

  Checking, $(7)^2 = 49$ and $(-7)^2 = 49$

- ✔ Solve for m in $3m^2 + 4 = 52$

  This isn't ready for the square root rule. Add –4 to each side.

  $$3m^2 = 48$$

  Now divide each side by 3.

  $$m^2 = 16$$

  So $m = \pm \sqrt{16} = \pm 4$

- ✔ Solve for p in $p^2 + 11 = 7$

  Add –11 to each side to get $p^2 = -4$.

Oops! What number times itself is equal to –4? The answer is: "None that you can imagine!" Actually, mathematicians have created numbers that don't actually exist so that these problems can be finished. The numbers are called *imaginary* numbers, but this book is concerned with the less heady numbers. So, this problem doesn't have an answer, if you're looking for a real number.

✔ Solve for q in $(q + 3)^2 = 25$

In this case, you end up with two completely different answers, not one number and its opposite.

Use the square root rule, first, to get $q + 3 = \pm \sqrt{25} = \pm 5$.

Now you have two different linear equations to solve:

$q + 3 = +5$ and $q + 3 = -5$

Subtracting 3 from each side of each equation, the two answers are: $q = 2$ and $q = -8$.

This one definitely needs to be checked.

Putting in the 2, $(2 + 3)^2 = 25$ or $5^2 = 25$.

Putting in the –8, $(-8 + 3)^2 = 25$ or $(-5)^2 = 25$.

Yes, they both work!

Factoring for a Solution

Here is where running through all the factoring methods can really pay off. In most quadratic equations, factoring is used rather than the square root rule method covered in the previous section. The square root rule is used when $b = 0$ in the quadratic equation $ax^2 + bx + c = 0$. Factoring is used when $c = 0$ or when neither b nor c is zero. A very important property used along with the factoring to solve these equations is the multiplication property of zero (MPZ). This is a very straightforward rule; it makes sense. Make use of the greatest common factor and the MPZ when solving the following equations.

Using the multiplication property of zero

Before you get into factoring, though, you need to know about the multiplication property of zero. You may say, "What's there to know? Zero multiplies anything and leaves nothing. It wipes out everything!" True enough, but there's this other nice property that zero has that is the basis of much equation solving in algebra. By itself, zero is nothing. Put it as the result of a multiplication problem, and you really have something: the multiplication property of zero.

The multiplication property of zero (MPZ) states that if $p \cdot q = 0$, then either $p = 0$ or $q = 0$.

This may seem obvious, but think about it. No other number has such a power over all other numbers. If you say that $p \cdot q = 12$, you can't predict a thing about p or q alone. These variables could be any number at all — positive, negative, fractional, radical, or a mixture of these. A product of zero, however, leads to one conclusion: One of the multipliers *must* be zero. No other means of arriving at a zero product exists.

The following examples show you how powerful and convenient this rule is.

✔ Find the value of x if $3x = 0$.

Use the multiplication property of zero. $x = 0$ because 3 can't be 0.

✔ Find the value of x and y if $xy = 0$.

If $x = 0$, then y can be any number, even zero. If $x \neq 0$, then y *must* be 0.

This property helps in solving quadratic equations provided that you move everything to one side, factor that side, and apply the multiplication property of zero. For example,

✔ Solve for x in $x^2 = 16$.

You could find the square root of each side to solve, but look at this alternate method first:

Subtract 16 from each side to get $x^2 - 16 = 0$.

The difference of two squares equals the sum of their roots times the difference of their roots.

Because $a^2 - b^2 = (a - b)(a + b)$

So $x^2 - 16 = (x - 4)(x + 4) = 0$

Using the multiplication property of zero:

$x - 4 = 0$ or $x + 4 = 0$

Solving those two linear equations, you find that $x = 4$ or $x = -4$.

When solving quadratic equations, because you're dealing with squares and second powers, it wouldn't hurt to have your list of perfect squares handy — use the one on the Cheat Sheet, if you like.

FUN FACT

Getting the quadratic second-degree

The word *quadratic* is used to describe equations that have a second-degree term. Why, then, is the prefix quad-, which means "four," used in a second-degree equation? It seems that this came about because a square is the regular four-sided figure, whose sides are the same. The area of a square with sides *x* long would be *x*-squared. So "squaring" in this case is raising to the second power.

Using the acronyms: GCF and MPZ

Often factoring is relatively simple when there are only two terms, and they have a common factor. This is true in quadratic equations of the form $ax^2 + bx + c = 0$ where $c = 0$. The two terms left will have the common factor of *x*, at least. You find the greatest common factor and factor that out, and then use the multiplication property of zero (MPZ) to solve the equation.

These examples make use of the fact that the constant term is zero, and there's a common factor of *x* or more.

⮑ Solve for *x* in $x^2 - 7x = 0$.

The greatest common factor of the two terms is *x*, so write the left side in factored form.

$$x(x - 7) = 0$$

Use the MPZ to say that $x = 0$ or $x - 7 = 0$, which gives you the two solutions $x = 0$ and $x = 7$.

⮑ Solve for *y* in $6x^2 + 18x = 0$.

The greatest common factor of the two terms is $6x$, so write the left side in factored form.

$$6x(x + 3) = 0$$

Use the MPZ to say that $6x = 0$ or $x + 3 = 0$, which gives you the two solutions $x = 0$ or $x = -3$.

⮑ Since $c = 0$ in so many quadratic equations, it might be useful to have a rule or formula for what the solutions are every time. Solve for *x* in this general quadratic equation where $c = 0$: $ax^2 + bx = 0$.

The greatest common factor of the two terms is *x*, so write the left side in factored form.

$$x(ax + b) = 0$$

Use the MPZ to say that $x = 0$ or $ax + b = 0$.

The first part of this is pretty clear. And this $x = 0$ business seems to crop up every time. The second part takes careful solving of the linear equation.

Subtract b from each side.

$$ax + b - b = 0 - b$$

$$ax = -b$$

Divide each side by a

$$\frac{ax}{a} = \frac{-b}{a} \text{ to get } x = \frac{-b}{a}$$

So the two solutions are $x = 0$ and $x = \frac{-b}{a}$.

This recognizable pattern can help you solve these types of equations. You can use this as a formula and not have to do the factoring and solving each time.

Missing the $x = 0$, the easiest part of these solutions, is amazingly easy. You see the lonely little x in the front of the parenthesis and forget that it gives you one of the two answers. Be careful.

Solving Quadratics with Three Terms

Quadratic equations are basic not only to algebra, but also to physics, business, astronomy, and many other applications. By solving a quadratic equation, you get answers to questions such as, "When will the rock hit the ground?" or "When will the profit be greater than 100 percent?" or "When, during the year, will the earth be closest to the sun?"

In the two previous sections, either b or c have been equal to zero in the quadratic equation $ax^2 + bx + c = 0$. Now I won't let anyone skip out. In this section, each of the letters, a, b, and c is a number that is *not* zero.

To solve a quadratic equation, moving everything to one side with zero on the other side of the equals sign is the most efficient method. Factor it if possible, and use the multiplication property of zero after you factor. If there aren't three terms, then refer to the previous sections.

Solve for x in $x^2 - 3x = 28$.

To solve three-term quadratics, follow these steps:

1. **Move all the terms to one side. Get zero alone on the right side.**

    In this case, you can subtract 28 from each side.

    $$x^2 - 3x - 28 = 0$$

Remember the standard form for a quadratic equation: $ax^2 + bx + c = 0$.

2. Determine all the ways you can multiply two numbers to get *a*.

In $x^2 - 3x - 28 = 0$, $a = 1$, which can only be 1 times itself.

3. Determine all the ways you can multiply two numbers to get *c*.

28 can be $1 \cdot 28$, $2 \cdot 14$, or $4 \cdot 7$

1. **If *c* is positive, find an operation from your Step 2 list and an operation from your Step 3 list that match so that the *sum* of their cross-products is the same as *b*.**

2. **If *c* is negative, find an operation from your Step 2 list and an operation from your Step 3 list that match so that the *difference* of their cross-products is the same as *b*.**

c is negative, and the difference of 4 and 7 is 3.

Factoring, you get $(x - 7)(x + 4) = 0$.

4. Use the multiplication property of zero (MPZ).

Either $x - 7 = 0$ or $x + 4 = 0$; now try solving for x by getting x alone to one side of the equal sign.

$x - 7 + 7 = 0 + 7$ or $x + 4 - 4 = 0 - 4$

which means that $x = 7$ or $x = -4$.

5. Check your answer.

If $x = 7$, then $(7)^2 - 3(7) = 49 - 21 = 28$, and

if $x = -4$, then $(-4)^2 - 3(-4) = 16 + 12 = 28$

They both check.

Factoring to solve quadratics sounds pretty simple on the surface. But factoring trinomial equations — those with three terms — can be a bit less simple. If a quadratic with three terms can be factored, then the product of two binomials is that trinomial. If the quadratic equation with three terms can't be factored, then use the quadratic formula (see "Figuring Out the Quadratic Formula" later in this chapter).

The product of the two binomials $(ax + b)(cx + d)$ is equal to the trinomial $acx^2 + (ad + bc)x + bd$.

This is a fancy way of showing what you get from using FOIL when multiplying the two binomials together. Now, on to using unFOIL. If you need more of a review of FOIL and unFOIL, check out Chapter 10.

The examples that follow all show how factoring and the MPZ allow you to find the solutions of a quadratic equation with all three terms showing.

Solve for x in $x^2 - 5x - 6 = 0$.

1. The equation is in standard form, so you can proceed.

2. Determine all the ways you can multiply to get a.

   $a = 1$, which can only be 1 times itself. If there are two binomials that the left side factors into, then they must each start with an x because the first term is x^2.

   $(x \quad)(x \quad) = 0$

3. Determine all the ways you can multiply to get c.

   $c = -6$, so the two binomials will end with either:
   +2 and –3 or –2 and +3 or +1 and –6 or –1 and +6.

   To decide which combination should be used, look at the last term in the trinomial, the 6, which is negative. This tells you that you have to use the *difference* of the absolute value of the two numbers in the list (think of the numbers without their signs) to get the middle term in the trinomial, the –5. In this case, one of the 1 and 6 combinations work, because their difference is 5. If you use the +1 and –6, then you get the –5 immediately from the cross product in the FOIL process.

   So $(x - 6)(x + 1) = 0$.

4. Use the multiplication property of zero.

   Using the MPZ, $x - 6 = 0$ or $x + 1 = 0$. This tells you that $x = 6$ or $x = -1$.

5. Check.

   If $x = 6$, then $(6)^2 - 5(6) - 6 = 36 - 30 - 6 = 0$.

   If $x = -1$, then $(-1)^2 - 5(-1) - 6 = 1 + 5 - 6 = 0$.

They both work!

Solve for x in $6x^2 + x = 12$.

1. Put the equation in the standard form.

   The first thing to do is to add –12 to each side to get the equation into the standard form for factoring and solving.

   $6x^2 + x - 12 = 0$

   This one will be a bit more complicated to factor because the 6 in the front has a couple of choices, and the 12 at the end has several choices. The trick is to pick the correct combination of choices.

2. Find all the combinations that can be multiplied to get a.

   6 can be obtained with a 1 and 6 or a 2 and 3.

3. Find all the combinations that can be multiplied to get c.

   12 can be obtained with a 1 and 12, or a 2 and 6, or a 3 and 4.

   You have to choose the factors to use so that the *difference* of their cross products (Outer and Inner) is 1. How do you know this? Because the 12 is negative, in this standard form, and the value multiplying the middle term is assumed to be 1 when there's nothing showing.

   Looking this over, you can see that using the 2 and 3 from the 6 and the 3 and 4 from the 12 will work: $2 \times 4 = 8$ and $3 \times 3 = 9$. The difference between the 8 and the 9 is, of course 1. You can worry about the sign later.

   I'll fill in the binomials and line up the factors so that the 2 multiplies the 4 and the 3 multiplies the 3 and I get a 6 in the front and 12 at the end. Whew!

   $$(2x \quad 3)(3x \quad 4) = 0$$

   The quadratic has a $+1x$ in the middle, so I need the *bigger* product of the Outer and Inner to be positive. I get this by making the $9x$ positive which happens when the 3 is positive and the 4 is negative.

   $$(2x + 3)(3x - 4) = 0$$

4. Use the MPZ to solve the equation.

   The trinomial has been factored. The multiplication property of zero tells you that either $2x + 3 = 0$ or $3x - 4 = 0$.

   If $2x + 3 = 0$ then $2x = -3$ or $x = -\frac{3}{2}$.

   If $3x - 4 = 0$ then $3x = 4$ or $x = \frac{4}{3}$.

5. Check your work.

   When $x = -\frac{3}{2}$, then $6\left(-\frac{3}{2}\right)^2 + \left(-\frac{3}{2}\right) = 12$

   $$6\left(\frac{9}{4}\right) - \frac{3}{2} = \frac{27}{2} - \frac{3}{2} = \frac{24}{2} = 12$$

   When $x = \frac{4}{3}$, then $6\left(\frac{4}{3}\right)^2 + \left(\frac{4}{3}\right) = 12$

   $$6\left(\frac{16}{9}\right) + \frac{4}{3} = \frac{32}{3} + \frac{4}{3} = \frac{36}{3} = 12$$

This checking wasn't nearly as fun as some, but it sure does show how well this factoring business can work.

Solve for y in $9y^2 - 12y + 4 = 0$.

1. This is already in the standard form.

2. Find all the numbers that multiply to get a.

   The factors for the 9 are 1×9 or 3×3.

3. Find all the numbers that multiply to get *c*.

   The factors for *c* are 1×4 or 2×2. Using the 3s and the 2s is what works because both cross products are 6, and you need a *sum* of 12 in the middle.

   So $9y^2 - 12y + 4 = (3y - 2)(3y - 2) = 0$. Notice that I put the negative signs in because the 12 needs to be a negative sum.

4. Use the MPZ to solve the equation.

   The two factors are the same here. That means that using the MPZ property gives you the same answer twice. When $3y - 2 = 0$, solve this for *y*. First add the 2 to each side, and then divide by 3. The solution is $y = \frac{2}{3}$. This is a double root, which, technically, has only one solution, but it occurs twice.

A *double root* occurs in quadratic equations that are perfect square binomials. These binomials are discussed in Chapter 8, if you need a refresher. These perfect square binomials are no more than the result of multiplying a binomial times itself. That's why, when they're factored, there's only one answer – it's the same one for each binomial.

Solve for *z* in $12z^2 - 4z - 8 = 0$.

1. This is already in standard form.

   You can start out by looking for combinations of factors for the 12 and the 8, but you may notice that all three terms are divisible by 4. To make things easier, take out that greatest common factor first, and then work with smaller numbers.

   $$12x^2 - 4z - 8 = 4(3z^2 - z - 2) = 0$$

2. Find the numbers that multiply to get *a*.

   $$a = 1 \times 3$$

3. Find the numbers to multiply to get *c*.

   This is really wonderful, especially because the 3 and 2 are both prime and can be factored only one way. Your only chore is to line up the factors so there will be a *difference* of 1 between them.

   $$4(3z^2 - z - 2) = 4(3z \quad 2)(z \quad 1) = 0$$

   Since the middle term is negative, you need to make the outer product negative, so put the negative sign on the 1.

   $$4(3z + 2)(z - 1) = 0$$

4. Use the MPZ to solve for the value of *z*.

   This time, when you use the MPZ, there are *three* factors to consider. Either $4 = 0$ or $3z + 2 = 0$ or $z - 1 = 0$. The first equation is impossible; 4 doesn't ever equal 0. But the other two equations give you answers. If $3z + 2 = 0$, then $z = -\frac{2}{3}$. If $z - 1 = 0$, then $z = 1$.

5. Checking:

If $z = -\frac{2}{3}$, then $12\left(-\frac{2}{3}\right)^2 - 4\left(-\frac{2}{3}\right) - 8 = 0$

and $12\left(\frac{4}{9}\right) + \frac{8}{3} - 8 = \frac{16}{3} + \frac{8}{3} - 8 = \frac{24}{3} - 8 = 8 - 8 = 0$

If $z = 1$, then $12(1)^2 - 4(1) - 8 = 12 - 4 - 8 = 0$

Notice that you use the original equation to check your solution — the equation before you did anything to it.

Applying Quadratic Solutions

Look at some of the ways to use quadratic equations and their solutions. Quadratic equations are found in many mathematics, science, and business applications; that's why they're studied so much. Here are some examples.

In physics, an equation that tells you how high an object is after a certain amount of time can be written: $h = -16t^2 + v_0 t + h_0$. In this equation, the $-16t^2$ part is the result of the pull of gravity on the object. The v_0 is the initial velocity — what the speed is at the very beginning. The h_0 is the starting height — the height in feet of the building, cliff, or stool from which the object is thrown. The variable t represents how many seconds or minutes have passed.

A stone was thrown upward from the top of a 40 foot building with a beginning speed of 128 feet per second. When was the stone 296 feet up in the air?

Replacing the height, h, with the 296 and the v_0 with 128 and the h_0 with 40, the equation now reads: $296 = -16t^2 + 128t + 40$. You can solve it using the following steps.

1. **Put the equation in standard form.**

   Add –296 to each side.

   $$0 = -16t^2 + 128t - 256$$

2. **Factor out the greatest common factor.**

   In this case, it's –16.

   $$0 = -16\left(t^2 - 8t + 16\right)$$

3. **Factor the quadratic.**

   $$0 = -16(t - 4)^2$$

4. **Use the MPZ to solve for the variable.**

   $$t - 4 = 0, \; t = 4$$

   After 4 seconds, the stone will be 296 feet up in the air.

This next example gets into the business side of these equations. The cost per pound of a special modeling clay is based on how many pounds a sculptor orders at one time. The formula for the cost per pound is $0.05x^2 - 4x + 100$ where x is the number of pounds of clay.

✔ Let y represent the cost per pound and write the equation $y = 0.05x^2 - 4x + 100$.

Replace the x in the equation with some numbers to get a handle on how much money is involved here.

For 1 pound, $x = 1$, and $y = 0.05(1)^2 - 4(1) + 100 = \$96.05$ per pound.

For 10 pounds, $x = 10$, and $y = 0.05(10)^2 - 4(10) + 100 = \$65.00$ per pound.

For 25 pounds, $x = 25$, and $y = 0.05(25)^2 - 4(25) + 100 = \$31.25$ per pound.

The sculptor wants to keep expenses down, and this means not spending more than \$40 per pound. How much can she order and keep it below this figure?

Let $y = 40$ and get the equation: $40 = 0.05x^2 - 4x + 100$

Subtracting 40 from each side, $0 = 0.05x^2 - 4x + 60$.

You don't want to have to factor this with the decimal multiplier of x^2, so multiply all the terms on both sides of the equal sign through by 20.

$$0 = x^2 - 80x + 1200$$

This factors into: $0 = (x - 60)(x - 20)$

Using the MPZ, $x = 60$ or $x = 20$.

The price per pound will stay below \$40 per pound if she orders between 20 and 60 pounds. As it turns out, it's the cheapest when she orders 40 pounds.

Figuring Out the Quadratic Formula

The quadratic formula is special to quadratic equations. A quadratic equation, written in its standard form, $ax^2 + bx + c = 0$, can have as many as two solutions, but there may be only one solution or even no solution at all. Remember, a, b, and c are any real numbers. The a can't equal zero, but the b or c can equal zero. The quadratic formula allows you to find solutions when the equations aren't very nice. The numbers in the equation or the solutions to the equation may be nasty fractions, decimals with no end, or radicals.

The *quadratic formula* says that if a quadratic equation is in the form $ax^2 + bx + c = 0$, then its solutions, the values of x, can be found with the equation

$$x = \frac{-b \pm \sqrt{b^2 - 4ac}}{2a}$$

There is an operation symbol ± in the formula. That means that the equation can be broken into two separate equations, one using the plus sign and the other using the minus sign. They look like the following:

$$x = \frac{-b + \sqrt{b^2 - 4ac}}{2a}$$

$$\text{or } x = \frac{-b - \sqrt{b^2 - 4ac}}{2a}$$

Can you see the difference between the two equations? The only difference is the change from the plus sign to the minus sign before the radical.

You can apply this formula to any quadratic equation to find the solutions. Here's an example of how the formula works.

> ✔ Use the quadratic formula to solve $2x^2 + 7x - 4 = 0$.
>
> Refer to the standard form of a quadratic equation where the coefficients of x^2 and x are a and b, and where the constant is c.
>
> $a = 2$, $b = 7$, $c = -4$
>
> Using the first part of the formula, with the plus in front of the radical,
>
> $$x = \frac{-7 + \sqrt{7^2 - 4(2)(-4)}}{2(2)}$$
> $$= \frac{-7 + \sqrt{49 + 32}}{4}$$
> $$= \frac{-7 + \sqrt{81}}{4}$$
> $$= \frac{-7 + 9}{4} = \frac{2}{4} = \frac{1}{2}$$
>
> Now, using the second part of the formula,
>
> $$x = \frac{-7 - \sqrt{7^2 - 4(2)(-4)}}{2(2)}$$
> $$= \frac{-7 - \sqrt{49 + 32}}{4}$$
> $$= \frac{-7 - \sqrt{81}}{4}$$
> $$= \frac{-7 - 9}{4} = \frac{-16}{4} = -4$$

Whenever the answers you get when using the quadratic formula come out as integers or fractions, it means that the equation *could have* been factored. It doesn't mean, though, that you shouldn't use the quadratic formula. Sometimes it's easier to use the formula if the equation has really large or nasty numbers. In general, though, it's quicker to factor using unFOIL and then the multiplication property of zero when you can. Just to illustrate this, look at the previous example when it's solved using factoring and the MPZ.

$$2x^2 + 7x - 4 = (2x - 1)(x + 4) = 0$$

Then using the multiplication property of zero to get:

$$2x - 1 = 0 \text{ or } x + 4 = 0$$
$$\text{so } x = \frac{1}{2} \text{ or } x = -4$$

So, what do the results look like when the equation *can't* be factored? The next example shows you.

The two things to watch out for when doing these problems are:

✔ Don't forget that −b means to use the *opposite* of b. If the coefficient b in the standard form of the equation is a positive number, then change it to the opposite, or the additive inverse, which would be a negative number, when using the formula. If b is negative, then change it to positive in the formula.

✔ Be careful when simplifying under the radical. The order of operations dictates that you square the value of b first, and then multiply the last three factors together before subtracting them from the square of b. Some sign errors can occur if you're not careful.

Try another example:

✔ Solve for x using the quadratic formula in $2x^2 + 8x + 7 = 0$.

In this problem, $a = 2$, $b = 8$, $c = -7$.

The formula is $x = \dfrac{-b \pm \sqrt{b^2 - 4ac}}{2a}$, but do the two solutions separately one more time.

$$x = \frac{-8 + \sqrt{8^2 - 4(2)(7)}}{2(2)}$$
$$= \frac{-8 + \sqrt{64 - 56}}{4} = \frac{-8 + \sqrt{8}}{4}$$

The radical can be simplified because $\sqrt{8} = \sqrt{4} \cdot \sqrt{2} = 2\sqrt{2}$,

$$\text{So } x = \frac{-8 + 2\sqrt{2}}{4} = \frac{2(-4 + \sqrt{2})}{4} = \frac{-4 + \sqrt{2}}{2}$$

Be careful when simplifying this expression: $\dfrac{(-4 + \sqrt{2})}{2} \neq -2 + \sqrt{2}$. Both terms in the numerator of the fraction have to be divided by the 2.

The decimal equivalent of the answer is:

$$\frac{-4 + \sqrt{2}}{2} \approx \frac{-4 + 1.414}{2} = \frac{-2.586}{2} = -1.293$$

The other solution is $x = \dfrac{-8 - \sqrt{8^2 - 4(2)(7)}}{2(2)} = \dfrac{-8 - \sqrt{64 - 56}}{4} = \dfrac{-8 - \sqrt{8}}{4}$

So $x = \dfrac{-8 - 2\sqrt{2}}{4} = \dfrac{2(-4 - \sqrt{2})}{4} = \dfrac{-4 - \sqrt{2}}{2}$

$\approx \dfrac{-4 - 1.414}{2} = \dfrac{-5.414}{2} = -2.707$

When you check these answers, what do the estimates do?

If $x = -1.293$, then $2(-1.293)^2 + 8(-1.293) + 7$

$= 3.343698 - 10.344 + 7 = -0.000302$

That isn't zero! What happened? Is the answer wrong? No, it's okay. The rounding did it. This is going to happen when you use a rounded value for the answer, rather than the radical form. An estimate was used for the $\sqrt{2}$ because it's an irrational number, and the decimal never ends. Rounding the decimal value to three decimal places seemed like enough decimal places. You shouldn't expect the check to come out to be exactly zero. In general, if you round the number you get from your check to the same number of places that you rounded your estimate of the radical, you should get the zero you're aiming for.

Chapter 15

Distinguishing Equations with Distinctive Powers

Most algebra applications involve solving equations of the first and second degree. Even in calculus and physics, these equations with powers of one and two seem to be enough to get through most of the applications. Do *these* well, and you'll do well. But every once in a while, you'll be thrown a curve with an equation of a degree higher than two or an equation with a radical in it or a fractional degree in it. No need to panic. You can deal with these many ways, and in this chapter, I tell you what those ways are. One common thread is that the equation is usually set equal to zero (except, sometimes, with radicals), so you use the multiplication property of zero.

Queuing Up to Cubic Equations

Cubic equations contain a term with a power of three but no power higher than three. In these equations, you can expect to find as many as *three* solutions, but there may not be three. Also, a cubic equation must have at least one solution, even though it may not be a nice one. Quadratic equations (a second-degree equation with a term that has an exponent of two) doesn't offer this guarantee: Quadratic equations don't have to have a solution.

If second-degree equations can have as many as two different solutions and third-degree equations can have as many as three different solutions, do you suppose that a pattern exists? Can you assume that fourth-degree equations could have as many as four solutions and fifth-degree equations . . . ? Yes, indeed you can — this is the general rule. The degree can tell you what the maximum number of solutions is. While the number of solutions may be less than the number of the degree, there won't be any *more* solutions than the number of the degree.

Solving perfectly cubed equations

If a cubic equation has just two terms and they're both perfect cubes, then your task is easy because the *sum* or *difference* of perfect cubes can be factored into two factors with only one solution. The first factor, or the binomial, gives you a solution. The second factor, the trinomial, does not give you a solution. (If you can't remember how to factor these things, go back to Chapter 11.)

If $x^3 - a^3 = 0$, then $x^3 - a^3 = (x-a)(x^2 + ax + a^2) = 0$ and $x = a$ is the only solution.

Look at how this works. If $(x-a)(x^2 + ax + a^2) = 0$, then the MPZ (the multiplication property of zero) says that either $x - a = 0$, which gives you the solution $x = a$, or $x^2 + ax + a^2 = 0$. This trinomial can't be factored and won't give you a solution. Likewise, if $(x+a)(x^2 - ax + a^2) = 0$, then the MPZ says that either $x + a = 0$, which gives you the solution $x = -a$, or $x^2 - ax + a^2 = 0$. This trinomial can't be factored and won't give you a solution. As you see, this rule uses the multiplication property of zero (MPZ), setting the binomial factor equal to 0 and solving for x.

The multiplication property of zero (MPZ) says that if $p \cdot q = 0$, then either $p = 0$ or $q = 0$. One of them has to be zero to get a product of zero.

The key to solving cubic equations that have two terms that are both cubes is in recognizing that that's what you have. Refer to the list of cubes on the Cheat Sheet. Then apply the rules. Here are some examples:

Solve for x in $x^3 - 8 = 0$

1. **Factor first.**

   The factorization is $x^3 - 8 = (x-2)(x^2 + 2x + 4)$.

2. **Apply the multiplication property of zero.**

   If $(x-2)(x^2 + 2x + 4) = 0$, then

   $x - 2 = 0$ or $x^2 + 2x + 4 = 0$.

Only the first equation, $x - 2 = 0$, has an answer: $x = 2$. The other equation doesn't have any numbers that satisfy it. There's only the one solution.

TECHNICAL STUFF

The trinomial that comes from factoring the sum or difference of two cubes can't be factored and doesn't have a solution when it's set equal to zero. To show that this is true, you can use the quadratic formula on the trinomial. This process is found in Chapter 14. Using the quadratic formula, you'd quickly find that there's no answer — at least not in the real world — that works in the equation.

- Solve for y in $27y^3 + 64 = 0$

  The factorization here is $27y^3 + 64 = (3y + 4)(9y^2 - 12y + 16)$.

  The first factor offers a solution, so set $3y + 4$ equal to zero to get $3y = -4$ or $y = -\frac{4}{3}$.

- Solve for a in $8a^3 - (a - 2)^3 = 0$

  The factorization here works the same as factorizations of the difference between perfect cubes. It's just more complicated because the second term is a binomial.

  $$8a^3 - (a - 2)^3 = \left[2a - (a - 2)\right]\left[4a^2 + 2a(a - 2) + (a - 2)^2\right] = 0$$

  Simplify inside the first bracket by distributing the negative and you get:

  $$\left[2a - (a - 2)\right] = \left[2a - a + 2\right].$$

  Simplify this to get $[a + 2]$.

  Simplifying inside the second bracket by distributing the $2a$ over the terms in the parenthesis and squaring the last binomial you get:

  $$\left[4a^2 + 2a(a - 2) + (a - 2^2)\right] = \left[4a^2 + 2a^2 - 4a + a^2 - 4a + 4\right]$$

  Simplifying, by combining like terms you get: $\left[7a^2 - 8a + 4\right]$

  $$\text{So } 8a^3 - (a - 2)^3 = [a + 2]\left[7a^2 - 8a + 4\right]$$

  Setting the first factor equal to 0, you get $a + 2 = 0$ or $a = -2$. As usual, the second factor doesn't give you a real solution.

Going for the greatest common factor

Another type of cubic equation that's easy to solve is one in which you can factor out a variable GCF (greatest common factor), leaving a second factor that is linear or quadratic (first or second degree). You apply the multiplication property of zero and work to find the solutions — usually three of them.

Factoring out a first-degree variable GCF

When a cubic equation has terms that all have the same first-degree variable as a factor, then factor that out. The resulting equation will have the variable as one factor and a quadratic expression as the second factor. The first-degree variable will always give you a solution of zero when you apply the MPZ. The quadratic can have solutions that you can find using the methods in Chapter 14.

Follow these steps to solve for x in $x^3 - 4x^2 - 5x = 0$

1. **Determine that each term has a factor of x and factor that out.**

   The GCF is x. Factor to get $x(x^2 - 4x - 5) = 0$.

   You're all ready to apply the MPZ when you notice that the second factor, the quadratic, can be factored. Do that first and then use the MPZ on the whole thing.

2. **Factor the quadratic expression, if possible.**

   $$x(x^2 - 4x - 5) = x(x - 5)(x + 1) = 0$$

3. **Apply the MPZ and solve.**

   Setting the individual factors equal to zero, you get

   $$x = 0 \text{ or } x - 5 = 0 \text{ or } x + 1 = 0.$$

   This means that $x = 0$ or $x = 5$ or $x = -1$.

4. **Check the solutions in the original equation.**

   If $x = 0$, then $0^3 - 4(0)^2 - 5(0) = 0 - 0 - 0 = 0$

   If $x = 5$, then $5^3 - 4(5)^2 - 5(5) = 125 - 4(25) - 25 =$

   $125 - 100 - 25 = 0$

   If $x = -1$, then $(-1)^3 - 4(-1)^2 - 5(-1) = -1 - 4(1) + 5 =$

   $-1 - 4 + 5 = 0$

All three work!

To solve for z in $z^3 + z^2 + z = 0$, follow the same steps:

1. Determine that each term has a factor of z and factor that out.

   Again, there's a common factor, and this time it's z.

   Factoring the z out, the equation reads $z(z^2 + z + 1) = 0$.

2. Factor the quadratic, if possible.

   This is where you get stuck. Even though factoring that second factor *looks* possible, factoring it is impossible. In fact, a solution is nonexistent when the second factor is set equal to 0. So the only solution is $z = 0$.

Factoring out a second-degree variable GCF

Just as with first-degree variable greatest common factors, you can also factor out second-degree variables. Factoring leaves you with a binomial that has a solution of the cubic equation. Just follow these easy dance steps:

1. **The steps for the rhumba are similar to the waltz: The rhythm is *slow-quick-quick* (one step followed by two quicker steps).**

2. **Step around on the left in a half turn.**

3. **Step directly into lead — the same for the waltz.**

Oops! I meant just follow these *factoring* steps to solve for w in $w^3 - 3w^2 = 0$:

1. Determine that each term has a factor of w^2 and factor that out.

   Find the common factor of w^2, and factor it out to get
   $w^3 - 3w = w^2(w - 3) = 0$.

2. Use the multiplication property of zero.

   $w^2 = 0$ or $w - 3 = 0$.

3. Solve the resulting equations.

   Solving the first equation involves taking the square root of each side of the equation. This process usually results in two different answers, the positive answer and the negative answer. However, this isn't the case with $w^2 = 0$ because 0 is neither positive nor negative. So there's only one solution from this factor: $w = 0$. And the other factor gives you a solution of $w = 3$. So, even though this is a cubic equation, there are only two solutions to it.

The following example gives you more practice with this process:

Solve for t in $9t^3 + 99t^2 = 0$

1. Factor out the greatest common factor.

   The greatest common factor of the two terms is $9t^2$. Factor to get

   $9t^3 + 99t^2 = 9t^2(t + 11) = 0$.

2. Solve the equation using the MPZ.

   Either $9t^2 = 0$ or $t + 11 = 0$. This means that $t = 0$ or $t = -11$.

Grouping cubes

Grouping is a form of factoring that you can use when you have four or more terms that don't have a greatest common factor. These four or more terms may be *grouped,* however, when two pairs of terms have factors in common.

The method of grouping is covered in Chapter 9. I give you one example here, but turn to Chapter 9 for a more complete review.

Solve for x in $x^3 + x^2 - 4x - 4 = 0$

1. **Use grouping to factor the left side.**

$$x^3 + x^2 - 4x - 4 = x^2(x+1) - 4(x+1) = (x+1)(x^2-4) = 0$$

2. **Factor the expressions in the parenthesis, if possible.**

   The second factor can also be factored because it's the difference between two perfect squares.

$$(x+1)(x^2-4) = (x+1)(x-2)(x+2) = 0$$

3. **Solve using the multiplication property of zero.**

   Use the MPZ. Either $x + 1 = 0$ or $x - 2 = 0$ or $x + 2 = 0$, which means that $x = -1$ or $x = 2$ or $x = -2$.

There are three different answers in this case, but you sometimes get just one or two answers.

Factoring cubics with integers

If you can't solve a third-degree equation by finding the sum or difference of the cubes, factoring, or grouping (see Chapter 9), you can try one more method *if* all the solutions happen to be integers. There could be one or two or three different integers that are solutions to a cubic equation. This generally only happens if the coefficient (multiplier) on the third-degree term is one. Just because the multiplier on the term to the third power is one doesn't necessarily mean that the answers are integers, but it's more likely to be so if that's the case. If the coefficient on the term with the variable raised to the third power isn't a one, then the solutions may be fractions. Synthetic division, which is covered in the "Dividing Synthetically" section later in this chapter, can be used to look for solutions.

To find the solutions when there are all integer solutions, follow these steps:

1. **Write the cubic equation in decreasing powers of the variable. Look for the *constant* (the term without a variable) and list all the numbers that divide that number evenly.**

   Remember to include both positive and negative numbers.

   Say you're solving for x in $x^3 - 7x^2 + 7x + 15 = 0$. The constant is 15, and the list of numbers that divides it evenly is: $\pm 1, \pm 3, \pm 5, \pm 15$. This is a long list, but you know that somehow or another you have to multiply numbers to get 15.

2. **Find a number from the list that makes the equation equal 0.**

   Choose a 3 for your first guess.

   Trying $x = 3$, $(3)^3 - 7(3)^2 + 7(3) + 15 = 27 - 63 + 21 + 15 = 63 - 63 = 0$.

   It works!

3. **Divide the constant by that number.**

   The answer to that division is your new constant.

   In the example, divide the original 15 by 3 and get 5. That's your new constant.

4. **Make a list of numbers that divide the new constant evenly.**

   Make a new list for the new constant of 5. The numbers that divide 5 evenly are: $\pm 1, \pm 5$. Two numbers are much nicer than four.

5. **Find a number from the new list that checks (makes the equation equal 0).**

   Trying $x = 1$, you get $(1)^3 - 7(1)^2 + 7(1) + 15 = 1 - 7 + 7 + 15 = 23 - 7 = 16$.

   That doesn't work, so try another number from the list.

   Trying $x = 5$, $(5)^3 - 7(5)^2 + 7(5) + 15 = 125 - 175 + 35 + 15 = 175 - 175 = 0$.

   So, it works.

6. **Divide the new constant by the newest answer.**

   That answer gives you the choices for the last solution.

   Dividing the new constant of 5 by 5, you get 1. The only things that divide that evenly are 1 or –1. Because you already tried the 1, and it didn't work, it must mean that the –1 is the last solution.

   Trying $x = -1$, you get $(-1)^3 - 7(-1)^2 + 7(-1) + 15 = -1 - 7 - 7 + 15 = 0$.

   That does work, of course, so your solutions are: $x = 3$, $x = 5$, $x = -1$.

This means that the three solutions to the equation $x^3 - 7x^2 + 7x + 15 = 0$ are 3, 5, and –1.

Whew! That's quite a process. Try it again with the following examples.

Solve for y in $y^3 - 4y^2 + 5y - 2 = 0$.

1. Write the equation in decreasing powers of the variable and find the constant term. Make a list of the numbers that divide the constant evenly.

   The constant in this one is –2. There's just a short list for this one: $\pm 1, \pm 2$.

2. Find a number from the list that makes the equation equal zero.

   Trying $y = 1$, $(1)^3 - 4(1)^2 + 5(1) - 2 = 1 - 4 + 5 - 2 = 6 - 6 = 0$

   This worked. The only disadvantage is that when you try to make the constant smaller, it doesn't work in this case. Dividing by 1 doesn't change the value. At least, it's a short list. Try another.

3. Try another number.

   Trying $y = -1$, $(-1)^3 - 4(-1)^2 + 5(-1) - 2 = -1 - 4 - 5 - 2 = -1 - 11 = -12$

   This one didn't work, so try a 2.

   Trying $y = 2$, $(2)^3 - 4(2)^2 + 5(2) - 2 = 8 - 16 + 10 - 2 = 18 - 18 = 0$

   The 2 works. So, if you divide the constant 2 by this 2, you get 1. The only factors for 1 are ± 1. You already tried 1, and it worked. You tried -1, and it didn't work. This means that the 1 will work again, and you have a double root of 1 in this problem.

   The solutions are $x = 1$, $x = 2$.

The way a double root or double solution works in these equations is that the solutions appear twice in the factored form. If you go backwards from the multiplication property of zero and write the factors that give the solutions to the cubic equation, it looks like this:

$$y^3 - 4y^2 + 5y - 2 = (y - 1)(y - 1)(y - 2) = 0$$

Or, showing the double root or solution more distinctly:

$$(y - 1)(y - 1)(y - 2) = (y - 1)^2 (y - 2) = 0.$$

Solving for z in $z^3 - 3z^2 + 3z - 1 = 0$ is special because it *could* have as many as three roots, but, because the constant is -1, the only choices are 1 or -1. Either there is only one solution, or one of them has to be repeated. To determine which situation you have, go through the steps and try values.

Trying $z = 1$, I get $(1)^3 - 3(1)^2 + 3(1) - 1 = 1 - 3 + 3 - 1 = 4 - 4 = 0$

Trying $z = -1$, I get $(-1)^3 - 3(-1)^2 + 3(-1) - 1 = -1 - 3 - 3 - 1 = -8$

The only solution is $z = 1$.

Working Quadratic-Like Equations

Some equations with higher powers or fractional powers are *quadratic-like*, meaning that they have three terms and

1. The first term has an even power (4, 6, 8, . . .) or $\left(\frac{1}{2}, \frac{1}{4}, \frac{1}{6}, \ldots\right)$.
2. The second term has a power that is half that of the first.
3. The third term is a constant number.

In general, the format is: $ax^{2n} + bx^n + c = 0$. Just as in the general quadratic equation, the x is the variable, the a, b, and c are constant numbers. The a can't be zero, but the other two letters have no restrictions. The n is also a constant and can be anything except zero. For example, if $n = 3$, then the equation would read $ax^6 + bx^3 + c = 0$.

To solve a quadratic-like equation, pretend that it is quadratic, and use the same method as you do for those — plus a step or two more. The extra steps usually involve taking an extra root or raising to an extra power.

Notice that each of the following quadratic-like equations meet all the requirements:

- $x^4 - 5x^2 + 4 = 0$
- $y^6 + 7y^3 - 8 = 0$
- $z^8 + 7x^4 + 6 = 0$
- $w^{1/2} - 7w^{1/4} + 12 = 0$

When you recognize that you have a *quadratic-like* equation, solve it by following these steps:

1. **Rewrite the quadratic-like equation as an actual quadratic equation, replacing the actual powers with two and one.**

   It makes sense to change the letters of the variables so that you don't confuse the rewritten equation with the original.

2. **Factor the new equation to see what the pattern is.**

3. **Use that pattern to factor the original problem, using the original variables.**

4. **Use the multiplication property of zero to find the solutions.**

The highest power of an equation, when it's a whole number, tells you the number of *possible* solutions; there won't be more than that number.

To put the solving steps to work, solve for x in $x^4 - 5x^2 + 4 = 0$.

1. **Rewrite the equation, replacing the actual powers with the numbers two and one.**

   Rewrite this as a quadratic equation using the same coefficients (number multipliers) and constant. Change the letter used for the variable, so you won't confuse this new equation with the original.

   $q^2 - 5q + 4 = 0$

2. **Factor the quadratic equation.**

   $q^2 - 5q + 4 = 0$ factors nicely into $(q - 4)(q - 1) = 0$

3. **Use this factorization pattern to factor the original equation.**

   Use that same pattern to factor the original problem. When you replace the variable q in the factored form, use x^2. You'll usually use that power in the middle when doing the factoring.

   $x^4 - 5x^2 + 4 = (x^2 - 4)(x^2 - 1) = 0$

4. **Solve the equation using the multiplication property of zero.**

   Now, use the MPZ. Either $x^2 - 4 = 0$ or $x^2 - 1 = 0$.

   If $x^2 - 4 = 0$, then $x^2 = 4$ and $x = \pm 2$.

   If $x^2 - 1 = 0$, then $x^2 = 1$ and $x = \pm 1$.

This equation did live up to its reputation and have four different solutions.

This next example presents an interesting problem because the exponents are fractions. But this fits into the category of quadratic-like, so I'll show you how you can take advantage of this situation to solve the equation. And, no, the rule of the number of solutions doesn't work the same way here. There aren't any possible situations where there's half a solution.

Solve $w^{1/2} - 7w^{1/4} + 12 = 0$.

1. Rewrite the equation with powers of two and one.

   Rewrite this to get $q^2 - 7q + 12 = 0$.

2. Factor.

   This factors nicely into $(q - 3)(q - 4) = 0$.

3. Replace the variables from the original equation, using the pattern.

   Replace with the original variables to get $(w^{1/4} - 3)(w^{1/4} - 4) = 0$.

4. Solve the equation.

   $(w^{1/4} - 3)(w^{1/4} - 4) = 0$

   Now, when you use the multiplication property of zero, you get that either $w^{1/4} - 3 = 0$ or $w^{1/4} - 4 = 0$. How do you solve these things?

Look at $w^{1/4} - 3 = 0$. Adding 3 to each side, you get $w^{1/4} = 3$. You can solve for w if you raise each side to the fourth power. That gets rid of the fractional exponent because when you raise a power to a power you multiply exponents. $\left(\frac{1}{4}\right)4 = 1$.

$$\left(w^{1/4}\right)^4 = (3)^4$$

This says that $w = 81$.

Doing the same with the other factor, if $w^{1/4} - 4 = 0$, then $w^{1/4} = 4$ and

$$\left(w^{1/4}\right)^4 = (4)^4$$

This says that $w = 256$.

5. Check the answers.

If $w = 81$, $(81)^{1/2} - 7(81)^{1/4} + 12 = 9 - 7(3) + 12 = 21 - 21 = 0$

If $w = 256$, $(256)^{1/2} - 7(256)^{1/4} + 12 = 16 - 7(4) + 12 = 28 - 28 = 0$

They both work.

Fractional exponents represent roots or radicals. So $x^{1/2} = \sqrt{x}$ and $x^{1/3} = \sqrt[3]{x}$. If you are given $16^{\frac{1}{4}}$, then that means $\sqrt[4]{16}$ or "What number multiplied by itself four times gives you 16?" In this case, the answer is 2.

Negative exponents are another interesting twist to these equations, as you see when you solve for the value of x in $2x^{-6} - x^{-3} - 3 = 0$.

1. **Rewrite the equation using powers of two and one.**

   Rewrite the equation as $2q^2 - q - 3 = 0$.

2. **Factor.**

   This factors into $(2q - 3)(q + 1) = 0$.

3. **Go back to the original variables and powers.**

   Use this pattern. Factor the original equation to get:

   $$\left(2x^{-3} - 3\right)\left(x^{-3} + 1\right) = 0$$

4. **Solve.**

   Use the MPZ. The two equations to solve are $2x^{-3} - 3 = 0$ and $x^{-3} + 1 = 0$.

   These become $2x^{-3} = 3$ and $x^{-3} = -1$.

   Rewrite these using the definition of negative exponents.

   $$x^{-n} = \frac{1}{x^n}$$

   So the two equations can be written $\frac{2}{x^3} = 3$ and $\frac{1}{x^3} = -1$

   Cross-multiply in each case and get $3x^3 = 2$ and $x^3 = -1$.

   Divide the first equation through by 3 to get the x^3 alone, and then take the *cube root* of each side to solve for x.

   $$x = \sqrt[3]{2/3} \text{ or } x = \sqrt[3]{-1} = -1$$

Physical challenges

One of the most famous musical composers was Beethoven. His accomplishments were even more incredible when you realize that he was deaf for a good deal of his life and still continued to produce musical masterpieces.

A similar situation occurred with the mathematician Leonhard Euler. Euler was one of the most prolific mathematicians of his generation and produced more than half of his work after he had gone blind. He dictated his findings from memory.

Rooting Out Radicals

Some equations have radicals in them. These equations get changed to linear or quadratic equations for greater convenience when solving. Radical equations crop up when doing problems involving distance in graphing points and lines. Also, problems involving the Pythagorean Theorem – that favorite of Pythagoras that describes the relationship between the sides of a right triangle. The basic process that leads to a solution is getting rid of the radical. Removing the radical changes the problem into something more manageable but introduces the possibility of a nonsense answer or an error. Checking your answer is even more important in the case of radicals. As long as you're aware that errors can happen, then you can be especially watchful. Even though this may seem a bit of a problem — that these nonsense things come up — getting rid of the radical is still the most efficient and easiest way to handle these equations.

Squaring both sides

The main method to use when dealing with equations that contain radicals is to change the equations to those that do *not* have radicals in them. You accomplish this by raising the radical to a power that changes the exponent to a one. If the radical is a square root, which can be written as a power of one-half, the radical is raised to the second power. If the radical is a cube root, which can be written as a power of one-third, then the radical is raised to the third power. Go back to Chapter 4 if you need to review exponents and raising to powers.

When the fractional power is raised to the reciprocal of that power, the two exponents are multiplied together giving you a power of one.

$$\left(\sqrt{2}\right)^2 = \left(2^{1/2}\right)^2 = 2^1 = 2$$

That works fine when you have only numbers. But what about

$$\left(\sqrt{x+1}\right)^2 = \left((x+1)^{1/2}\right)^2 = (x+1)^1 = x+1?$$

This works fine, too, but problems can occur when variables are used instead of numbers. Variables can stand for negative numbers or values that allow negatives under the radical, which isn't always apparent until you get into the problem and check an answer.

The equations containing radicals can be dealt with by squaring or raising each side to an appropriate power. Remember the adage: "Do unto one side of the equation what you do unto the other."

To work a radical equation (or even a conservative equation containing radical terms), follow these steps:

1. **Get the radical by itself on one side of the equal sign.**

   So, if you're solving for y in $\sqrt{4-5y} - 7 = 0$, add 7 to each side to get the radical by itself on the left. Doing that gives you $\sqrt{4-5y} = 7$.

2. **Square both sides of the equation to remove the radical.**

   Squaring both sides of the example problem gives you $4 - 5y = 49$.

3. **Solve the resulting linear equation.**

   Working through the linear equation gives you $-5y = 45$ or $y = -9$.

   It may seem strange that the answer is a negative number, but the negative number is multiplied by another negative, which makes the answer a positive number. Then you add 4 under the radical.

4. **Check your answer.**

   If $y = -9$ then $\sqrt{4-5(-9)} - 7 = 0$ or $\sqrt{4+45} - 7 = 0$. That leads to $\sqrt{49} - 7 = 7 - 7 = 0$. It checks!

You can use the same method to solve for x in $2\sqrt{x+15} - 3 = 9$.

1. Get the radical term by itself on one side of the equation.

   The first step is to add 3 to each side:

   $$2\sqrt{x+15} = 12$$

2. Square both sides of the equation.

   One of the rules involving exponents is that squaring the product of two whole numbers is equal to the product of each of those same whole numbers squared.

   $$(a \cdot b)^2 = a^2 \cdot b^2$$

   Squaring the left side, $\left(2\sqrt{x+15}\right)^2 = 2^2 \cdot \left(\sqrt{x+15}\right)^2 = 4(x+15)$

Squaring the right side, $12^2 = 144$

$$4(x+15) = 144.$$

3. Solve for x.

Distribute $4x + 60 = 144$

Subtract 60 from each side.

$$4x = 84 \text{ or } x = 21.$$

4. Check your work.

$$2\sqrt{x+15} - 3 = 9$$
$$2\sqrt{21+15} - 3 = 2\sqrt{36} - 3 = 2 \cdot 6 - 3 = 12 - 3 = 9$$

Try your hand at this slightly more complicated problem: Solve for z in $7 + \sqrt{z-1} = z$.

1. Get the radical by itself on the left.

Subtracting 7 from each side, you end up with the radical on the left and a binomial on the right.

$$\sqrt{z-1} = z - 7$$

2. Square both sides of the equation.

The only thing to watch out for here is squaring the binomial correctly.

$$\left(\sqrt{z-1}\right)^2 = (z-7)^2$$
$$z - 1 = z^2 - 14z + 49$$

3. Solve the equation.

This time you have a quadratic equation. Move everything over to the right, so that you can set the equation equal to 0. To do this, subtract z from each side and add 1 to each side.

$$0 = z^2 - 15z + 50$$

This factors into $(z-5)(z-10) = 0$. Using the multiplication property of zero, you get either $z - 5 = 0$ or $z - 10 = 0$, which in turn means that either $z = 5$ or $z = 10$.

4. Check.

Check these carefully because impossible answers often show up in these problems.

If $z = 5$, then $7 + \sqrt{5-1} = 7 + \sqrt{4} = 7 + 2 = 9 \neq 5$. The 5 doesn't work.

If $z = 10$, then $7 + \sqrt{10-1} = 7 + \sqrt{9} = 7 + 3 = 10$. The 10 *does* work.

The only solution is that z equals 10. That's fine. Sometimes these problems have two answers, sometimes just one answer, or sometimes no answer at all. The method works; you just have to be careful.

Squaring both sides twice

Just when you thought things couldn't get any better, up comes a situation where you have to square both sides of an equation not once, but twice! This happens when you have more than one radical in an equation and getting it alone on one side of the equation isn't possible.

As you go about solving the problem, a binomial is formed with two radicals or a radical and another term, and you have to square terms twice to get rid of all the radicals. The procedure is a little involved, but nothing too horrible. You can find out how to go about things in the following steps:

Solve for the value of x in the equation $\sqrt{x-3} + 4\sqrt{x+6} = 12$ by following these steps:

1. **Get the radicals on either side of the equals sign.**

   Even though you can't get either radical by itself, having them on either side of the equation helps. So subtract the $\sqrt{x-3}$ from each side to put it on the right with the 12: $4\sqrt{x+6} = 12 - \sqrt{x-3}$.

2. **Square both sides of the equation.**

   On the left side, this involves the rule involving exponents where you're squaring a product. This rule is covered in Chapter 4 and in the section right before this one. On the right side, this involves squaring a binomial (using FOIL), which gives you:

   $$\left(4\sqrt{x+6}\right)^2 = \left(12 - \sqrt{x-3}\right)^2$$
   $$16(x+6) = 144 - 24\sqrt{x-3} + \left(\sqrt{x-3}\right)^2$$
   $$16x + 96 = 144 - 24\sqrt{x-3} + x - 3$$

3. **Simplify, and get the remaining radical by itself on one side of the equation.**

   Simplifying in the example means doing the math on the constants — 144 and –3, to get:

   $$16x + 96 = 141 - 24\sqrt{x-3} + x$$

   Subtract 141 and x from each side:

   $$15x - 45 = -24\sqrt{x-3}$$

4. **Look for a common factor in all the terms of the equation.**

   You can make things a bit neater by dividing each side by the greatest common factor, 3:

   $$\cancel{3}(5x - 15) = \cancel{3}\left(-8\sqrt{x-3}\right)$$
   $$5x - 15 = -8\sqrt{x-3}$$

   Now you can square both sides more easily:

5. **Square both sides of the equation.**

   $$(5x - 15)^2 = \left(-8\sqrt{x-3}\right)^2$$
   $$25x^2 - 150x + 225 = 64(x-3)$$
   $$25x^2 - 150x + 225 = 64x - 192$$

   These are some rather large numbers.

6. **Get everything on one side of the equation and factor.**

   You can move everything to the left and see whether you can factor anything out to make the numbers smaller. In the example, you can subtract $64x$ from each side and add 192 to each side.

   $$25x^2 - 214x + 417 = 0$$

   This isn't the easiest quadratic to factor, but it does factor into $(25x - 139)(x - 3) = 0$. So, you have two solutions. Either $x = \dfrac{139}{25}$ or $x = 3$.

7. **Plug in the solutions to check your answer.**

   If $x = \dfrac{139}{25}$, then $\sqrt{139/25 - 3} + 4\sqrt{139/25 + 6} = 12$. What are the chances of this? You can get out your trusty calculator to see if it works.
   $\sqrt{64/25} + 4\sqrt{289/25} = 8/5 + 4(17/5) = 76/5 \neq 12$. After all that, the answer doesn't even work! Hope for the 3.

   If $x = 3$, then $\sqrt{3-3} + 4\sqrt{3+6} = 0 + 4\sqrt{9} = 4 \cdot 3 = 12$. Oh, good!

Dividing Synthetically

Cubic equations that have nice, integer solutions make life easier. But how realistic is that? Many nice answers are fractions. And what if you want to broaden your horizons and try fourth- or fifth-degree equations or higher? Trying answers until you find one that works can get pretty old pretty fast.

A method known as synthetic division can help out with all these concerns. The division looks a little strange — it's synthesized. *Synthesize* means to bring together separate parts. That's what a synthesizer does with music. So turn on the Beethoven and get going.

Simple synthesizing

Synthetic division is a short-cut division process. It takes the coefficients on all of the terms in an equation and provides a method for finding the answer to a division problem by only multiplying and adding. It's really rather neat.

Find the solutions to the cubic equation $x^3 + x^2 - 14x - 24 = 0$. Using the method involved in listing all the possible solutions — all the numbers that divide the constant evenly, you see that 24 has factors of $\pm 1, \pm 2, \pm 3, \pm 4, \pm 6, \pm 8, \pm 12, \pm 24$. This method is covered earlier in the chapter under "Factoring cubics with integers." Trying different choices could be quite a challenge here. Use synthetic division, instead. Write the coefficients (multipliers) of the terms in a row.

$$1 \qquad 1 \qquad -14 \qquad -24$$

Now put a backwards/upside down division sign in front and a line below the numbers. The backwards/upside down division sign is used to place dividing choices in front of the coefficients. This type symbol usually means that there is an unusual type of division going on.

$$\underline{|1 \qquad 1 \qquad -14 \qquad -24}$$

Next bring down the first 1.

$$\underline{|1 \qquad 1 \qquad -14 \qquad -24}$$
$$1$$

Now you're ready to *guess*. If a number from the list of choices is a solution of the equation, then the remainder, which will be in the last, right-most position in the synthetic division, is 0. The numbers in the bottom row represent the numbers in the quotient of the division — except the last number, which is the remainder. You want a 0 in the bottom row, beneath the –24. That means that you've found a solution.

Guess that 4 works. Put the 4 in the funny division symbol.

$$\underline{4|1 \qquad 1 \qquad -14 \qquad -24}$$
$$1$$

Now multiply, add, multiply, add, multiply, and add. Each time, you multiply the right-most number in the bottom row times the number in the division sign. Put the result above the line in the next space to the right, and add the two numbers together. The bottom 1 times the 4 equals 4, so put it under the second 1.

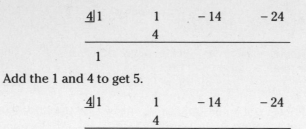

```
4|1        1      - 14      - 24
          4
_____
   1
```

Add the 1 and 4 to get 5.

```
4|1        1      - 14      - 24
          4
_____
   1      5
```

Multiply the 5 times the 4 and add the product to the –14.

```
4|1        1      - 14      - 24
          4        20
_____
   1      5        6
```

One more time with the 6 times 4 and add that product to the –24.

```
4|1        1      - 14      - 24
          4        20        24
_____
   1      5        6         0
```

Eureka! The 4 worked, so it's a solution. Now, to try the next guess, use the numbers in the bottom row (except the last 0). And the list of guesses has shrunk to just those that divide 6. This bottom row represents the quotient — or result of the division. By finding 4 as a solution, its factor has been eliminated from the process. Dividing the original constant, 24, by 4 you get a new constant, 6, and its set of numbers to divide by.

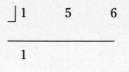

```
 |1        5        6
_____
  1
```

Try a 2 this time.

```
2|1        5        6
          2        14
_____
  1        7        20
```

This didn't work. The remainder isn't 0.

Try a –2.

```
-2|1        5        6
           - 2      - 6
_____
   1        3        0
```

This one worked.

The last division involves only the two numbers that are left.

The only choices here are 3 and –3. If you want a sum of 0, you'd better use the –3.

This worked, too, so your solutions are $x = 4, -2, -3$.

Synthetic division is so useful because it is easier to find multiple zeros and fractional answers. Fractions for answers come up when the multiplier on the first term (the term with the highest power) is something other than 1.

Synthesizing with fractions

An equation, such as $4x^3 - 3x^2 + 2x + 6 = 0$, can have as many as three solutions. And two of them may be fractions. Synthetic division is a good way to find these answers, but you need to start from the list of all possible solutions before starting the synthetic division.

The list of all possible solutions for any of these problems consists of:

1. All the numbers that divide the constant evenly.
2. All the numbers that divided the constant evenly *divided by* all the numbers that divide the coefficient of the term with the highest power evenly.

In the example $4x^3 - 3x^2 + 2x + 6 = 0$, the list consists of:

First, all the numbers that divide 6 evenly.

Second, all the numbers that divide 6 evenly *divided by* all the numbers that divide 4 evenly.

You can see why the problem is nicer when the coefficient of the highest power is a one.

The list of guesses starts as usual, with all the numbers that divide 6 evenly. So start with: $\pm 1, \pm 2, \pm 3, \pm 6$. The fractions are found by dividing each of the factors of 6 by the numbers that divide 4, the multiplier of the first term, evenly. The numbers that divide 4 evenly are: $\pm 1, \pm 2, \pm 4$.

If you divide $\pm 1, \pm 2, \pm 3, \pm 6$ by the first pair, ± 1, then nothing changes.

If you divide $\pm 1, \pm 2, \pm 3, \pm 6$ by the second pair, ± 2, then you get $\pm\frac{1}{2}, \pm\frac{2}{2}, \pm\frac{3}{2}, \pm\frac{6}{2} = \pm\frac{1}{2}, \pm 1, \pm\frac{3}{2}, \pm 3$. As you can see, there are some repeats from the original list.

If you divide $\pm 1, \pm 2, \pm 3, \pm 6$ by the third pair, ± 4, then you get $\pm\frac{1}{4}, \pm\frac{2}{4}, \pm\frac{3}{4}, \pm\frac{6}{4} = \pm\frac{1}{4}, \pm\frac{1}{2}, \pm\frac{3}{4}, \pm\frac{3}{2}$. Again, there are more repeats, but the list is complete. Any fractional answers are in those lists.

Chapter 16

Fixing Inequalities

. . . .

. . . .

Equality — a powerful word in social, political, and humanitarian arenas. And, it's no less powerful as far as mathematics is concerned: Algebra wouldn't have much without equality. Fortunately, algebra knows how to deal with inequality too (far better than in those other arenas). Equality is an important tool in mathematics and science. This chapter introduces you to algebraic inequality, which isn't exactly the opposite of equality. You could say that algebraic inequality is a bit like equality but softer. You use inequality for comparisons. Inequality is used when determining if something is positive or negative — bigger than or smaller than zero. Inequality allows you to sandwich expressions between values on the low end and the high end, such as the tides were between five feet and eighteen feet.

Algebraic inequalities show relationships between a number and an expression or between two expressions. One expression is bigger or smaller than another for certain values of a given variable. For instance, it could be that Janice has at least four more than twice as many cats as Eloise. There are lots of answers that can occur if it's *at least* and not exactly as many.

Equations, statements with equal signs, are one type of relation — two things are exactly the same, it says. The inequality relation is a bit less precise. One thing can be bigger by a lot or bigger by a little, but there's still that relationship between them — that one is bigger than the other.

Many operations on inequalities work the same as operations on equalities and equations, but you need to pay attention to some important differences.

Operating on Inequalities

There are many similarities between working with inequalities and working with equations. The balancing part still holds. It's when operations like multiplying each side by a number or dividing each side by a number come into play that there are some differences.

The inequality symbols are

- < **less than**
- > **greater than**
- ≤ **less than or equal to**
- ≥ **greater than or equal to**

Remember that the *little* point is next to the *smaller* of the two values.

The rules for operations on inequalities are:

If $a < b$, then $a + c < b + c$ and $a - c < b - c$.

If $a < b$ and c is positive, then $a \cdot c < b \cdot c$ and $\frac{a}{c} < \frac{b}{c}$.

If $a < b$ and c is negative, then $a \cdot c > b \cdot c$ and $\frac{a}{c} > \frac{b}{c}$.

Notice that the direction of the symbol has changed.

Adding and subtracting inequalities

Adding and subtracting values within inequalities works exactly the same as with equations. You keep things balanced.

Start with an inequality statement that you can tell is true by looking at it, such as six is less than ten: $6 < 10$.

What happens if you add the same thing to each side? You can do that to an equation and not have the truth change, but what about an inequality?

$$6 < 10$$

Add 4 to each side.

$$6 + 4 < 10 + 4$$

$$10 < 14$$

Ten is less than 14 is still a true statement. This demonstration isn't enough to *prove* anything, but it does illustrate a rule that is true. When you add any number to both sides of an inequality, the inequality is still correct or true.

When you subtract any number from both sides of an inequality, the inequality is still correct or true.

$10 < 14$

Subtract 2 from each side.

$10 - 2 < 14 - 2$

$8 < 12$

Eight *is* less then 12, so it looks as if adding and subtracting are okay. But you stayed with positive numbers and positive results. How about adding a negative number to each side that makes both sides negative?

$8 < 12$

Add –24 to each side.

$8 + (-24) < 12 + (-24)$

$-16 < -12$

This is still true: –16 is farther from 0 than –12.

Multiplying and dividing inequalities

Now come the tricky operations. Multiplication and division add a new dimension to working with inequalities.

When multiplying or dividing both sides of an inequality by a *positive* number, the inequality remains correct or true. When multiplying or dividing both sides by a *negative* number, the inequality sign has to be reversed — point in the opposite direction — for the inequality to be correct or true. You can never multiply each side by zero — that always makes it false. And, of course, you can never divide *anything* by zero.

Start with positive numbers, such as 20, which is greater than 12:

$20 > 12$

Multiply each side by 4.

$20 \cdot 4 > 12 \cdot 4$

$80 > 48$

It's still true. So is there a problem?

You can see the problem by starting with a new inequality, such as $10 > -3$, and multiplying each side by -2.

$10 > -3$

Multiply each side by -2

$10(-2) > -3(-2)$

$-20 > 6$

Oops! A negative can't be greater than a positive.

$-20 < 6$

Making the inequality untrue is bad news. The good news is that turning the sign around is a relatively easy way to fix this.

Whenever you multiply each side of an inequality by a *negative* number (or divide by a negative number), turn the inequality sign to face the opposite direction.

Multiplying or dividing by a positive number does *not* change the direction. But if you take $-6 < 12$ and multiply each side by -3, as in $-6(-3) > 12(-3)$, you turn the sign around: $18 > -36$. Turning the sign around makes the less-than sign a greater-than sign.

Now, for division, take $18 > -36$ and divide each side by -9: $\frac{18}{-9} < \frac{-36}{-9}$. Make sure to switch the inequality sign from a greater-than sign to a less-than sign: $-2 < 4$.

In the case of inequalities, you can neither divide nor multiply by zero. Of course, dividing by zero is always forbidden, but you can usually multiply expressions by zero (and get a product of zero). However, you cannot multiply inequalities by zero.

Look at what happens when each side of an inequality is multiplied by zero.

$3 < 7$

$0 \cdot 3 < 0 \cdot 7$

$0 < 0$

No! It's just not true: Zero is not less than itself, nor is it greater. So, to keep zero from getting an inferiority or superiority complex, don't use it to multiply inequalities.

Solving Linear Inequalities

Linear inequalities, like linear equations, are those in which the exponent on the variable is no more than one. Solving them is much like solving linear equations. The main thing to remember is to reverse the inequality symbol when you multiply or divide by a negative number — and only then. You also need to keep in mind that you don't get just a single answer but a whole bunch of answers — an *infinite* number of answers. These numbers that are *all* the answers to the inequality will have to meet some qualifications. For instance, they may all be larger than a certain number or smaller than a certain number or between two certain numbers. The answer or solution could be something like: x is bigger than three; all numbers bigger than three can replace the x and make the inequality a true statement.

Here are a few examples to demonstrate the similarities and differences between working with equations and inequalities.

1. **Move all variable terms to one side.**

   Solve for all the values of x that work in $4x - 2 < x + 10$

   Start this problem like an equation. Move the x variables to the left and the numbers to the right. First, subtract x from each side.

   $4x - x - 2 < x - x + 10$

   $3x - 2 < 10$

   Now add 2 to each side.

   $3x - 2 + 2 < 10 + 2$

   $3x < 12$

2. **Divide each side by the coefficient of the variable.**

   As long as the number you divide by is positive, you can leave the inequality pointing in the original direction. Divide by +3.

   $$\frac{3x}{3} < \frac{12}{3}$$

   $x < 4$

   The answer tells you that any number smaller than 4 makes a true statement in the original problem. Unfortunately, there's no single check that will do. You can try a few test numbers to be sure you haven't forgotten to turn the inequality sign around. Try some here.

3. Check the answer.

If the answers to $4x - 2 < x + 10$ are that $x < 4$, then the following should work:

If $x = 3$, then $4(3) - 2 < 3 + 10$, or $10 < 13$. This one works.

If $x = -10$, then $4(-10) - 2 < -10 + 10$, or $-42 < 0$. This one works, also.

In this next problem, you solve for the values of z in $-2(3z + 4) > 10$.

In this case, the variable term is already on the left. A usual next step would be to distribute the -2 over the terms on the left. But, because 2 divides 10 evenly, an alternate step lets you avoid having to do the distribution. This is a good option when the division doesn't result in any fractions; otherwise, you should go ahead and distribute.

1. Divide each side by -2.

$$\frac{-2(3z + 4)}{-2} < \frac{10}{-2}$$

$3z + 4 < -5$.

Notice that the inequality sign has been reversed.

2. Subtract 4 from each side.

$$3z + 4 - 4 < -5 - 4$$

$$3z < -9$$

3. Divide each side by 3.

$$\frac{3z}{3} < \frac{-9}{3}$$

$$z < -3$$

4. Check.

Let $z = -9$, then $-2([3(-9)] + 4) > 10$,
or $-2(-27 + 4) > 10$

$-2(-23) > 10$, or $46 > 10$

It checks.

Working with More Than Two Expressions

One big advantage that inequalities have over equations is that they can be expanded or strung out, and you can do more than one comparison at the same time. Look at this statement:

$2 < 4 < 7 < 11 < 12$

You can create another true statement by pulling out any pair of numbers from the inequality, as long as you write them in the same order. They don't even have to be next to one another. For example:

$4 < 12$ or $2 < 11$ or $2 < 12$

One thing you *cannot* do, though, is to mix up inequalities, going in *opposite* directions, in the same statement. You *cannot* write $7 < 12 > 2$.

The operations on these multi-inequality expressions use the same rules as for the linear expressions (refer to the "Operating on Inequalities" section earlier in this chapter). You just extend them to each section or part.

Here's the first statement: $2 < 4 < 7 < 11 < 12$

Add 5 to each section: $7 < 9 < 12 < 16 < 17$

Multiply each by –1, and reverse the inequality, of course:

$-7 > -9 > -12 > -16 > -17$

Take this info and solve for the values of x in $-3 \le 5x + 2 < 17$.

Notice that both the less-than-or-equal-to and the less-than signs are used in the same statement. That's allowed because they both face the same direction.

1. The goal is to get the variable alone in the middle. Start by subtracting 2 from *each section*.

   $-3 - 2 \le 5x + 2 - 2 < 17 - 2$

   $-5 \le 5x < 15$

2. Now divide each section by 5. This is positive, so don't turn the inequality signs around.

   $-1 \le x < 3$

   This says that x is bigger than or equal to –1 while, at the same time, it's smaller than 3. Some possible solutions are: 0, 1, 2, $-\frac{1}{2}$, 2.9.

3. Check the problem using two of these possibilities.

   If $x = 1$, then $-3 \le 5(1) + 2 < 17$, or $-3 \le 7 < 17$

   That's true.

   If $x = \frac{-1}{2}$, then $-3 \le 5\left(\frac{-1}{2}\right) + 2 < 17$, or $-3 \le \frac{-1}{2} < 17$.

   This also works.

Try solving for x in $16 > 1 - 3x > 7$.

1. Start by subtracting 1 from each section.

   $16 - 1 > 1 - 1 - 3x > 7 - 1$

   $15 > -3x > 6$

2. Now divide each section by –3. This means reversing each of the inequality symbols.

   $\frac{15}{-3} < \frac{-3x}{-3} < \frac{6}{-3}$

   $-5 < x < -2$

   This means that x has to be a number between –5 and –2. Some examples of numbers that would work are: –4, –3, –2.1.

3. Check.

   You see that if $x = -3$, then $16 > 1 - 3(-3) > 7$, or $16 > 10 > 7$.

   It seems to work!

Solving Quadratic Inequalities

A *quadratic inequality* is an inequality that involves a term with a second-degree power. The rules of addition, subtraction, multiplication, and division of inequalities still hold, but the final step in solving these is different. Working out these quadratic inequalities is almost like a puzzle that falls neatly into place as you work on it. The steps for dealing with this type of inequality follow:

Look at this quadratic inequality: $x^2 + 3x > 4$.

The answers to these inequalities can go in more than one direction — the numbers can be bigger than one number or smaller than another number — so I'm going to demonstrate that these things can work before showing you how to solve them.

Make some guesses as to what works for x in this expression.

- ✔ For instance, if $x = 2$, then $(2)^2 + 3(2)$ is 4 + 6; 10 > 4, so 2 works.

- ✔ Also, if $x = 5$, then $(5)^2 + 3(5)$ is 25 + 15; 40 > 4 so 5 works also. It looks like the bigger the better.

- ✔ If $x = 0$, then $(0)^2 + 3(0)$ is 0 + 0; 0 is *not* > 4, so, no, 0 doesn't work. How about negative numbers?

- ✔ Now, if $x = -6$, then $(-6)^2 + 3(-6)$ is 36 – 18; 18 > 4, so, yes, –6 works. Some negatives work; some positives work. There's a method you can use to find which work and which don't work without all this guessing.

To solve quadratic inequalities, follow these steps:

1. **Move all terms to one side so that the terms are greater than or less than zero.**

2. **Factor.**

3. **Find all values of the factored side that make that side equal to zero.**

4. **Make a chart listing the values, in order, that make the expression equal to zero. Leave spaces between the numbers for signs. Determine the signs (positive and negative) of the expression *between* those values that make it equal zero and write them on the chart.**

Try the preceding steps on the quadratic inequality $2x^2 + 5x \geq 12$.

1. Move all terms to one side.

    First, move the 12 to the left by subtracting 12 from each side.

    $$2x^2 + 5x - 12 \geq 12 - 12$$
    $$2x^2 + 5x - 12 \geq 0$$

2. Factor.

    Factor the quadratic on the left using unFOIL.

    $$(2x - 3)(x + 4) \geq 0$$

3. Find all the values of x that make the factored side equal to zero.

    In this case, there are two values. Using the multiplication property of zero, you get $2x - 3 = 0$ or $x + 4 = 0$, which results in $x = \frac{3}{2}$ or $x = -4$.

4. Make a chart listing the values from step 3, and determine the signs of the expression between the values on the chart.

    Look *between* $\frac{3}{2}$ and -4. The expression *equals* 0 when x is $\frac{3}{2}$ or -4. The expression is positive for all the numbers between the two that make it zero or it's negative for all numbers between the two that make it equal to zero. Just testing one of the numbers between tells you what will happen to *all* of them. First put these numbers in order from smaller to larger. That would put the -4 before the $\frac{3}{2}$.

5. Go back to the problem, after it's been changed to get 0 on one side, and try some numbers smaller than the smaller value for *x*, larger than the larger value for *x*, and numbers between those two values.

    In this case, you try numbers smaller than -4, between -4 and $\frac{3}{2}$, and larger than $\frac{3}{2}$.

    Use $2x^2 + 5x - 12 \geq 0$

First, choose some numbers smaller than –4.

If $x = -5$, then $2(-5)^2 + 5(-5) - 12 = 50 - 25 - 12 = +13$

If $x = -10$, then $2(-10)^2 + 5(-10) - 12 = 200 - 50 - 12 = +138$

Actually, no matter what number you choose that's smaller than –4, you get a positive number. So write that on the blank to the left of the –4.

$\underline{\text{Positive}} -4 \underline{\hspace{2cm}} \frac{3}{2} \underline{\hspace{2cm}}$

Now try some numbers between –4 and $\frac{3}{2}$.

If $x = -2$, then $2(-2)^2 + 5(-2) - 12 = 8 - 10 - 12 = -14$

If $x = 1$, then $2(1)^2 + 5(1) - 12 = 2 + 5 - 12 = -5$

No matter what number you choose between these two numbers, you'll get a negative number. So write that on the blank in the middle.

$\underline{\text{Positive}} -4 \underline{\text{Negative}} \frac{3}{2}$

Just one more place to check. Look at some numbers bigger than $\frac{3}{2}$.

If $x = 2$, then $2(2)^2 + 5(2) - 12 = 8 + 10 - 12 = 6$.

If $x = 10$, then $2(10)^2 + 5(10) - 12 = 200 + 50 - 12 = 238$.

No matter what number you choose bigger than $\frac{3}{2}$, you always get a positive number. So write that on the last blank.

$\underline{\text{Positive}} -4 \underline{\text{Negative}} \frac{3}{2} \underline{\text{Positive}}$

6. Look again at the inequality statement to figure out what numbers to plug in.

The sample problem is $2x^2 + 5x - 12 \geq 0$.

The inequality reads that the expression is greater than or equal to 0. Something that's *greater than* 0 is a positive number. So you're really interested in when the expression is positive. According to the chart, the expression is positive when x is smaller than –4, and then again, when x is bigger than $\frac{3}{2}$. Because you want the *or-equal-to* part, include the –4 and $\frac{3}{2}$.

7. Write the answer.

So, $x \leq -4$ or $x \geq \frac{3}{2}$.

The result for x is *many* numbers — all the negative numbers smaller than –4 way down to really small numbers and all the positive numbers bigger than $\frac{3}{2}$ way up to really big numbers. The only numbers that *don't* work are those between –4 and $\frac{3}{2}$.

In this example, you solve for the values of y in $y^2 + 8y + 15 < 0$.

1. All the terms are already on one side.

2. Factor.

$$(y + 3)(y + 5) < 0$$

3. Find the values of y that make the factored expression equal to zero.

   This time, even though the values of y that make the left side equal to 0 are not going to be a part of the solution, you still need them to find the answers. The numbers you want are –3 and –5.

4. Make a chart using the values that make the expression equal to zero.

   When $y = -3$ or $y = -5$, the left side of the expression equals zero. Put them in order with the smaller, then the larger. Leave blanks again, and fill them in with *positive* or *negative* signs.

   _____ –5 _____ –3 _____

5. Fill in the spaces between the numbers on the chart with the sign of the expression.

   Looking at the original expression, $y^2 + 8y + 15 < 0$, look at numbers in each of the three areas and fill in the blanks.

   If $y = -6$, then $(-6)^2 + 8(-6) + 15 = 36 - 48 + 15 = +3$: *Positive*

   If $y = -4$, then $(-4)^2 + 8(-4) + 15 = 16 - 32 + 15 = -1$: *Negative*

   If $y = 1$, then $(1)^2 + 8(1) + 15 = 1 + 8 + 15 = +24$: *Positive*

   <u>Positive</u> –5 <u>Negative</u> –3 <u>Positive</u>

   The original statement, $y^2 + 8y + 15 < 0$, is true when the value of the expression is less than 0, or negative. So the numbers between –5 and –3 are solutions of the inequality.

6. Write that:

   $$-5 < y < -3$$

This time it's just the numbers between; the solution doesn't include the –5 or –3.

Working without zeros

Sometimes the values that make the expression in an inequality equal to zero are nonexistent, which means that the expression never changes sign. It's always negative or always positive. You only have to determine whether anything solves the problem.

For example, the expression $x^2 + 4$ in the inequality $x^2 + 4 \geq 0$ doesn't factor. And any number you put in for x gives you a positive value on the left. So this statement is *always* positive, and the inequality is true for all numbers.

Dealing with more than two factors

Even though this section involves problems that are *quadratic inequalities* (an inequality that has at least one squared term and a greater-than or less-than sign), some other types of inequalities belong in the same section because you handle them the same way. You can really have any number of factors and any arrangement of factors and do the positive-and-negative business to get the answer, as the examples show.

✔ Solve for the values of x that work in $(x - 4)(x + 3)(x - 2)(x + 7) > 0$.

This problem is already factored, so you can easily determine that the numbers that make the expression equal to 0 are $x = 4$, $x = -3$, $x = 2$, $x = -7$.

Put them in order from the smallest to the largest and leave blanks between them.

_____ –7_____ –3 _____ 2 _____ 4 _____

This time, put the test values back into the factored form. This is easier than multiplying it all out. The numbers are going to get pretty big, anyway.

If $x = -8$, then $(-8 - 4)(-8 + 3)(-8 - 2)(-8 + 7) = (-12)(-5)(-10)(-1)$:

Positive.

This multiplies out to be + 600, but you really didn't need to know the number value. Just the fact that there are four negatives tells you that the product is positive.

When multiplying or dividing integers, if the number of negative signs in the problem is even, the result is positive. If the number of negative signs in the problem is odd, the result is negative. For example, in this problem:

If $x = -5$, then $(-5 - 4)(-5 + 3)(-5 - 2)(-5 + 7) = (-9)(-2)(-7)(+2)$: *Negative*

If $x = 0$, then $(0 - 4)(0 + 3)(0 - 2)(0 + 7) = (-4)(+3)(-2)(+7)$: *Positive*

If $x = 3$, then $(3 - 4)(3 + 3)(3 - 2)(3 + 7) = (-1)(+6)(+1)(+10)$: *Negative*

If $x = 5$, then $(5 - 4)(5 + 3)(5 - 2)(5 + 7) = (+1)(+8)(+3)(+12)$: *Positive*

Now, filling in the blanks,

Positive –7 Negative –3 Positive 2 Negative 4 Positive

Because the original problem is looking for values that make the expression greater than 0, or positive, then the solution includes numbers smaller than –7, between –3 and 2, and bigger than 4. It reads:

$x < -7$ or $-3 < x < 2$ or $x > 4$

Please look at one more example. In case you got the impression that these inequalities have nice positive-negative-positive patterns, this example shows you that you always need to take the time to check these out carefully.

✔ Solve for the values of x in $(x - 1)^2(3 - x)(x + 2) \le 0$.

The numbers making the expression equal to 0 are $x = 1, 3, -2$. Put them in order.

_____ –2 _____1 _____3 _____

Checking *between* the zeros:

If $x = -3$, then $(-3 - 1)^2(3 - [-3])(-3 + 2) = (+16)(+6)(-1) = -96$: *Negative*

If $x = 0$, then $(0 - 1)^2(3 - 0)(0 + 2) = (+1)(+3)(+2) = +6$: *Positive*

If $x = 2$, then $(2 - 1)^2(3 - 2)(2 + 2) = (+1)(+1)(+4) = +4$: *Positive*

If $x = 4$, then $(4 - 1)^2(3 - 4)(4 + 2) = (+9)(-1)(+6) = -54$: *Negative*

Filling in the blanks:

Negative –2 Positive 1 Positive 3 Negative .

In this case, you need variables (x) that make the expression negative or zero. You can see that values smaller than –2 work and values bigger than 3 work. So do all the 0s — the numbers separating the areas. Notice that the number 1 works, even though it's isolated between positive areas.

The answer reads:

$x \le -2$ or $x = 1$ or $x \ge 3$.

Figuring out fractional inequalities

Inequalities with fractions are another special type of inequality that fits under the general heading of quadratic inequalities because of the way you solve them. Try to do somewhat the same business as with the inequalities dealing with more than two factors by finding where the expression equals zero. Actually, look to see, separately, what makes the numerator (top) equal to zero and what makes the denominator (bottom) equal to zero. Check the areas between the zeros; and then write out the answer.

The one big caution with these is not to include any number in the answer that makes the denominator of the fraction equal zero. That makes it an impossible situation, not to mention an impossible fraction. So why look at what makes the denominator zero at all? The number zero separates positive numbers from negative numbers. Even though the zero itself can't be used in the solution, it indicates where the sign changes from positive to negative or negative to positive.

- Try to solve for the values of y in $\dfrac{(y+4)}{(y-3)} > 0$.

  Because the rules involving the signs of the answers in multiplication and division are the same (odd number of negative signs means a negative answer, an even number means a positive answer), you use the same procedure as with quadratics and more than two terms.

  The numbers making the numerator or the denominator equal to 0 are:

  $y = -4$ or $y = 3$

  Making a chart:

  _____ -4 _____ 3 _____ .

  If $y = -5$, then $\dfrac{(-5+4)}{(-5-3)} = \dfrac{-1}{-8} = +\dfrac{1}{8}$: *Positive*

  If $y = 0$, then $\dfrac{(0+4)}{(0-3)} = \dfrac{4}{-3} = \dfrac{-4}{3}$: *Negative*

  If $y = 4$, then $\dfrac{(4+4)}{(4-3)} = \dfrac{8}{1} = +8$: *Positive*

  Filling in the chart:

  Positive -4 Negative 3 Positive

  The problem only asks for values that make the expression greater than 0, or positive, so the solution is: $y < -4$ or $y > 3$.

- Solve for the values of z in $\dfrac{(z^2-1)}{(z^2-9)} \le 0$

  The numbers making the numerator or denominator equal to 0 are:

  $z = +1, -1, +3, -3$.

  The chart: _____ -3, _____ -1 _____ 1 _____ 3 _____

  If $z = -4$, then $\dfrac{([-4]^2-1)}{([-4]^2-9)} = +\dfrac{15}{7}$: *Positive*

If $z = -2$, then $\dfrac{\left([-2]^2 - 1\right)}{\left([-2]^2 - 9\right)} = -\dfrac{3}{5}$: *Negative*

If $z = 0$, then $\dfrac{\left([0]^2 - 1\right)}{\left([0]^2 - 9\right)} = +\dfrac{1}{9}$: *Positive*

If $z = 2$, then $\dfrac{\left([2]^2 - 1\right)}{\left([2]^2 - 9\right)} = -\dfrac{3}{5}$: *Negative*

If $z = 5$, then $\dfrac{\left([5]^2 - 1\right)}{\left([5]^2 - 9\right)} = +\dfrac{24}{16}$: *Positive*

Now the chart reads:

<u>Positive</u> –3 <u>Negative</u> –1 <u>Positive</u> 1 <u>Negative</u> 3 <u>Positive</u>.

The original expression is:

$$\frac{\left(z^2 - 1\right)}{\left(z^2 - 9\right)} \le 0$$

Because you're looking for values of z that make the expression negative, you want the values between –3 and –1 and those between 1 and 3. Also, you want values that make the expression equal to 0. That can only include the numbers that make the *numerator* equal to 0, the 1 and –1. The answer is written:

$$-3 < z \le -1 \text{ or } 1 \le z < 3.$$

Notice that the < symbol is used by the –3 and 3 so they don't get included in the answer.

Working with Absolute Value Inequalities

Absolute value inequalities are just what they say they are — inequalities that have absolute value symbols somewhere in the problem.

$|a|$ is equal to a if a is a positive number or 0. $|a|$ is equal to the opposite of a, $-a$, if a is a negative number.

So $|3| = 3$ and $|-4| = 4$.

Absolute value equations and inequalities can look like:

$$|x| = 7 \quad |2x + 3| > 7 \quad |5x + 1| \le 9$$

Working absolute value equations

Before tackling the inequalities, take a look at *absolute value equations*. An equation such as $|x| = 7$ is fairly easy to decipher. It's asking for values of x that give you a 7 when you put it in the absolute value symbol. Two answers, 7 and –7, have an absolute value of 7. But what about something a bit more involved, such as $|3x + 2| = 4$?

To solve an absolute value equation of the form $|ax + b| = c$, change the absolute value equation to two equivalent linear equations and solve them.

$|ax + b| = c$ is equivalent to $ax + b = c$ and $ax + b = -c$

Notice that the left side is the same in each case. The c is positive in the first equation and negative in the second because the expression in the absolute value symbol can be positive or negative — absolute value makes them both positives when it's performed.

Solve for x in $|3x + 2| = 4$.

1. **Rewrite as two linear equations.**

   $3x + 2 = 4$ or $3x + 2 = -4$

2. **Solve for the value of the variable in each of the equations.**

   Subtract 2 from each side in each equation:

   $3x = 2$ or $3x = -6$

   Divide *each* side in each equation by 3:

   $x = \dfrac{2}{3}$ or $x = -2$

3. **Check.**

   If $x = \dfrac{2}{3}$, then $\left| 3\left(\dfrac{2}{3}\right) + 2 \right| = |2 + 2| = |4| = 4$

   $x = -2$, then $|3(-2) + 2| = |-6 + 2| = |-4| = 4$

They both work.

Solving for x in $|5x - 2| + 3 = 0$, however, is one case where you *don't* want the equation set equal to zero. In order to use the rule for changing to linear equations, you have to have the absolute value by itself on one side of the equation.

1. **Get the absolute value expression by itself on one side of the equation.**

   Adding –3 to each side:

   $|5x - 2| = -3$

2. **Rewrite as two linear equations:**

$5x - 2 = -3$ or $5x - 2 = +3$

3. **Solve the two equations for the value of the variable.**

Add 2 to each side of the equations.

$5x = -1$ or $5x = 5$

Divide each side by 5:

$x = -\dfrac{1}{5}$ or $x = 1$

4. **Check.**

If $x = -\dfrac{1}{5}$

then, $\left| 5 \left(-\dfrac{1}{5} \right) - 2 \right| = \left| -1 - 2 \right| = 3$

Oops! That's supposed to be a –3. Try the other one.

If $x = 1$, then $\left| 5 \left(1 \right) - 2 \right| = \left| 1 - 2 \right| = 3$

No, that didn't work either.

Now is the time to realize that the equation was impossible to begin with. (Of course, noticing this before you started would have saved time.) The definition of absolute value tells you that it results in everything being *positive*. Starting with an absolute value equal to –3 gave you an impossible situation to solve. No wonder you didn't get an answer!

Working absolute value inequalities

Solving absolute value inequalities brings two different procedures together into one. The first involves the methods used to deal with absolute value, and the second involves the rules used to solve inequalities. You might say it's the best of both worlds. Or you might not.

To solve an absolute value inequality of the form $|ax + b| > c$, change the absolute value inequality to two equivalent inequalities and solve them.

$|ax + b| > c$ is equivalent to $ax + b > c$ or $ax + b < -c$

To solve an absolute value inequality of the form $|ax + b| < c$, change the absolute value inequality to an equivalent inequality and solve it.

$|ax + b| < c$ is equivalent to $-c < ax + b < c$

This first example involves *greater than* and uses the rule for changing the problem into something doable.

Solve for x in $|2x - 5| > 7$.

1. Rewrite as two inequalities.

$2x - 5 > 7$ or $2x - 5 < -7$

2. Solve each inequality.

Add 5 to each side in each inequality.

$2x > 12$ or $2x < -2$

Divide through by 2.

$x > 6$ or $x < -1$

This second example involves *less than* and uses the second part of the rule for changing the problem into something doable.

Solve for x in $|5x + 1| < 9$

1. Rewrite as two inequalities:

$-9 < 5x + 1 < 9$

2. Solve the inequality.

Subtract 1 from each section.

$-10 < 5x < 8$

Now divide through by 5.

$-2 < x < \dfrac{8}{5}$

Solve for x in $|5x - 2| + 1 < 9$.

1. Get the absolute part on one side by itself.

You can't apply the rule until you subtract 1 from each side.

$|5x - 2| < 8$

2. Rewrite.

$-8 < 5x - 2 \leq 8$

Add 2 to each section.

$-6 \leq 5x \leq 10$

And, now, divide by 5.

$-\dfrac{6}{5} \leq x \leq 2$

Inequality is a special type of relationship that has similarities to and differences from equality. The different rules for dealing with different types of problems aren't difficult, but you have to follow them faithfully and precisely.

Part IV
Applying Algebra

The 5th Wave   By Rich Tennant

"David's using algebra to calculate the tip. Barbara—would you mind being a fractional exponent?"

In this part . . .

No algebra book can ever be complete without talking about how to apply algebra to everyday situations. This part provides the motivation for doing all the preparation and work in the other parts. Formulas and story problems provide the most frequently used applications of algebra. Add a few graphs for good measure, and you have it all.

Chapter 17

Making Formulas Behave

I remember applying for my first car loan. What a traumatic experience! The application form was the first challenge. But the most mysterious and awe-inspiring part of the whole experience was when the loan officer sat across the desk with his calculator and started entering numbers. It seemed as if he pushed buttons for hours; he'd pause and reflect; he'd push more numbers and frown. And then he looked up, smiled, and said, "Yes. Approved." What formula was he using? I'll never know.

Most of life's formulas aren't nearly so scary. And formulas that are a bit complicated can be tamed with a little know-how and a decent dose of confidence. This chapter is all about gaining experience, know-how, and confidence.

When you use algebra in the real world, more often than not you turn to a formula to help you work through a problem. Fortunately, when it comes to algebraic formulas, you don't have to reinvent the wheel: You can make use of standard, tried-and-true formulas to solve some common, everyday problems.

Not all formulas you come across are in this chapter. But there should be enough explanation and practice on the ones that are here to get you through any that come your way. In this chapter, you walk through (or you work through?) some formulas that you can use again and again.

Measuring Up

Some universal concerns — some start at an early age — are those dealing with measurements. How far is it? How big are you? How much room do you need, anyway? How much more wrapping paper are you going to need? These questions all have to do with measurements and, usually, formulas.

Finding out how long: Units of length

Before measurements were standardized, they varied according to who was doing the measuring: A yard was the distance from the tip of the nose to the end of an outstretched arm; a foot, well, you can probably guess where that came from; and an inch was often the length of the second bone in the index finger. Perhaps the "perfect man" modeled for the measures used today.

The units of measure most commonly used in the United States are inches, feet, yards, and miles.

Some equivalent measures are: 12 inches = 1 foot; 3 feet = 1 yard; 5,280 feet = 1 mile.

You can change these equivalencies into formulas by saying:

✔ Feet to inches: number of inches = number of feet × 12

✔ Inches to feet: number of feet = number of inches ÷ 12.

✔ Yards to feet: number of feet = number of yards × 3

✔ Miles to feet: number of feet = number of miles × 5,280

The best way to deal with these and other measures is to write a proportion. To review the properties of proportions, see Chapter 13.

When using a proportion to solve a measurement problem, write inches over inches and feet over feet (same units over same units). For example:

✔ How many inches in 8 feet?

12 inches = 1 foot

That's what's known.

$$\frac{12 \text{ in.}}{x \text{ in.}} = \frac{1 \text{ ft.}}{8 \text{ ft.}}$$

The *problem* is on the bottom. Now cross-multiply.

$12 \times 8 = x \times 1$

$x = 96$ inches

Eight feet contain 96 inches.

✔ You're in a plane, and the pilot says that you're cruising at 14,000 feet. How high is that in miles?

$$5,280 \text{ feet} = 1 \text{ mile}$$

$$\frac{5,280 \text{ ft.}}{14,000 \text{ ft.}} = \frac{1 \text{ mi.}}{x \text{ mi.}}$$

Cross-multiply.

$$5,280 \times x = 14,000 \times 1$$

$$5,280x = 14,000$$

$$x = \frac{14,000}{5,280} = 2\frac{3,440}{5,280} \text{ mi.} \approx 2.65 \text{ mi. up in the air.}$$

Putting the Pythagorean theorem to work

Another great formula to use when working with lengths and distances is the Pythagorean theorem. The *Pythagorean theorem* is a formula that shows the special relationship between the three sides of a right triangle. A *right triangle* (as opposed to a *wrong* triangle?) is one with a 90-degree angle. Pythagoras noticed that if a triangle really was a right triangle, then the square of the length of the longest side (the *hypotenuse*) is always equal to the sum of the squares of the two shorter sides.

$$(\text{length of hypotenuse})^2 = (\text{length of other side})^2 + (\text{length of remaining side})^2$$

FUN FACT

Puzzling Pythagoras

Pythagoras was born somewhere around 570 B.C. He is most well known for his Pythagorean theorem, but he is also responsible for discovering an important musical property: The notes sounded by a vibrating string depend on the length of the string.

Pythagoras was a great thinker, but he also exhibited some rather bizarre behavior. He founded a school where about 300 young aristocrats studied mathematics, politics, philosophy, religion, music, and astronomy. They formed a very tight fraternity or secret society where the members had their diets and actions regulated. They weren't allowed to eat beans or drink wine or pick up anything that had fallen or stir a fire with an iron. They had to face in a certain direction when they urinated. These strange beliefs supposedly caused Pythagoras' death. When he was being chased from his burning home by some persecutors, he was supposed to have stopped at the edge of a bean field, and, rather than trample the beans, allowed his chasers to catch and kill him.

For instance, a triangle with sides measuring 3 inches, 4 inches, and 5 inches is a right triangle. The longest side is the one that measures 5 inches. The square of 5 is 25. The two shorter sides are 3 and 4 inches. $3^2 = 9$ and $4^2 = 16$. The sum of 9 and 16 is 25 — the square of the longest side. This works *only* for right triangles, and if it works, the triangle *has* to be a right triangle.

Pythagorean theorem: If *a, b,* and *c* are the lengths of the sides of a right triangle, and *c* is the longest side (the hypotenuse), then

$$a^2 + b^2 = c^2$$

The following examples show how you can use the Pythagorean theorem.

✔ Show that a triangle with sides that measure 5, 12, and 13 is a right triangle.

First, find the square of the measure of each side.

$5^2 = 25$, $12^2 = 144$, $13^2 = 169$

Add the two smaller squares together. That sum is the same as the largest square: $25 + 144 = 169$

✔ Show that a *multiple* of the sides measuring 3, 4, and 5 of a right triangle is also a right triangle.

Multiply 3, 4, and 5 each by 8. The sides are then 24, 32 and 40. Find the square of the measures of the sides.

$24^2 = 576$, $32^2 = 1024$, $40^2 = 1600$

$576 + 1024 = 1600$

This is a specific multiple, but any multiple will work.

✔ A carpenter wants to determine whether a garage doorway has square corners or if it's really leaning to one side. He measures 30 inches from one corner along the bottom of the doorway and makes a mark. He measures 40 inches up from the same corner and makes a mark on the side. He then takes a tape measure and measures the distance between the marks. It comes out to be 49 inches.

Find the squares of the measures.

$30^2 = 900$, $40^2 = 1600$, $49^2 = 2401$

$900 + 1600 = 2500 \neq 2401$

The two smaller squares *don't* add up to the larger square.

No, it's not square.

Of course, in carpentry the measures can't always come out exactly as the math says they should. But there are guidelines that carpenters use to tell when it's close enough.

Working around the perimeter

How long is the running track around the field? What's the distance around the room? How many feet of fencing do you need to go around the pool? The *perimeter* is the distance around the outside of a given area — the total length of the periphery that borders a region.

In general, the perimeter of a figure is the sum of the lengths of the sides. If you have a triangle, measure each of the three sides and add them up. If you have a four-sided figure, add up the four lengths, and so on. Perimeter formulas are used to simplify this process when you have special types of figures and you recognize that the figure is something special. You can use quick, easy formulas to do the computations. Many of these formulas are listed below.

Triangulating triangles

The perimeter of a triangle is equal to the sum of the measures of the three sides — sides 1, 2, and 3:

$$P = s_1 + s_2 + s_3$$

The formula for the perimeter of a triangle shows the variable *s* and the subscripts 1, 2, and 3. The "s" stands for side. Rather than use an *a*, *b*, and *c*, it's customary to use a single variable, like *s* in this case, and a number of subscripts when there's nothing special about the sides or how their lengths relate to one another. Subscripts are also used when there are more than twenty-six sides in a figure. You can go only *a* through *z* to name the sides, but with subscripts you can go on as long as you like — the figure could have a thousand sides. Heaven forbid!

- ✔ Find the perimeter of the triangle with sides 5 ft., 11 ft. and 13 ft.

    $P = 5 + 11 + 13 = 29$ft.

- ✔ Find the amount of fencing you'll need for a triangular area if the two sides that form a right triangle are 7 yards and 24 yards, and you can't measure the longest side, the hypotenuse, because it's too muddy right now.

    Because it's a right triangle, the sum of the squares of 7 and 24 is equal to the square of the longest side. $7^2 + 24^2 = 49 + 576 = 625$. 625 is the square of 25, so the sides of the area are 7, 24, and 25 yards.

    $P = 7 + 24 + 25 = 56$ yards of fencing needed.

Squaring up to squares

A square is wonderful to work with because you have only one measure to worry about — the length of one side is the same as all the others.

Eratosthenes prime

Eratosthenes of Cyrene was a mathematician who lived from about 275 to 195 B.C. He is famous for his work with prime numbers and is credited with being the first to estimate, accurately, the diameter of the earth.

He was the director of the famous library in Alexandria for several decades. Very little of his original work has survived. He died at about 80 years of age when he starved himself to death because he was so depressed at having gone blind.

The perimeter is 4 times the length of a side:

$$P = 4s$$

✔ An environmental group is going to search a square mile of prairie to check for toxins in beetles. What is the perimeter of that square mile in *feet?*

One mile is 5,280 feet. So the perimeter is

$$4 \cdot 5,280 = 21,120 \text{ feet}$$

So, if they want to rope off the area, they need plenty of rope!

Rectangles

A *rectangle* is a special four-sided figure. The first shape in Figure 17-1 is a rectangle with square (90-degree) corners, and the opposite sides are the same length.

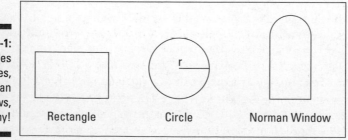

Figure 17-1: Rectangles and circles, and Norman windows, oh my!

Rectangle   Circle   Norman Window

Find the perimeter of a rectangle by adding up the lengths of the sides. Another way is to double the length and double the width and add the two numbers together. And, yet, another way is to add the length and width together and then multiply that sum by two.

$$P = s_1 + s_2 + s_3 + s_4$$

$$P = 2l + 2w = 2(l + w)$$

For instance, if the sides measure 6 inches, 9 inches, 6 inches, 9 inches, you can just add: 6 + 9 + 6 + 9 = 30. No big deal. But sometimes, when the numbers aren't as nice, it's easier to pair up these numbers. In rectangles, each measure of a side occurs twice. So just find the two sides that have different measures — they're next to one another — and add the two different numbers together and double the sum: 2(6 + 9) = 2(15) = 30. The *length* is usually the longer side and the *width* is usually the shorter side.

> ✔ Your new garden is a rectangle measuring 85 feet long by 35 feet wide. How much fencing do you need to enclose it? What's the perimeter?
>
> Add the 85 and 35 together and double it.
>
> > 2(85 + 35) = 2(120) = 240 feet of fencing
>
> Of course, this doesn't include a gate — you should probably consider that, too, unless you like jumping hurdles.

Circumscribing circles

A circle has a perimeter, but there's a special name for it: *circumference*. Think about the word: If the *circumstances* (conditions around you) are positive, you can *circumnavigate* (sail around), *circumvent* (go around and avoid), and *circumscribe* (draw around). To find the circumference of a circle, all you need is the measure of the radius. The *radius* is the distance from the center of the circle to any point on the circle. If you double the radius, you get the measure of the diameter. The second shape in Figure 17-1 shows a circle with the radius marked. The *diameter* is the length from one side of the circle to the other — through the center.

The formula for circumference (distance around the outside of a circle) is $C = \pi d$ where d is the *diameter* (distance across the circle) and π is always about 3.14.

The symbol for *is about* or *is approximately equal to* is written $\approx$. So $\pi \approx 3.14$.

π is the symbol for the Greek letter pi. This symbol traditionally stands for the relationship between the diameter (distance across) and the circumference (distance around) of a circle. The relationship is the same, no matter how large or small a circle is. The relationship is that the circumference is *always* a little more than three times as big as the diameter. The *little more* part can't be written exactly — the decimal value goes on forever — but usually the value of π is approximated using 3.14, 3.1416, or $\frac{22}{7}$. These are the most popular values. Which value you use depends on the circumstances. I usually use 3.14.

✔ You want to rewrite the formula so you can easily determine how wide your circular garden will be if you buy a bundle of fencing to put around it and use all the fencing in the bundle. The fencing comes in bundles of 50, 100, 150, or 200 feet, and so on.

Solving for d in the formula $C = \pi d$:

$\frac{C}{\pi} = \frac{\pi d}{\pi}$ Divide each side by π.

$d = \frac{C}{\pi}$

The diameter is equal to the circumference divided by π.

$d = \frac{C}{3.14}$

If the bundle has 50 feet of fencing,

$d = \frac{50}{3.14} \approx 15.92$ feet across

If the bundle has 100 feet of fencing,

$d = \frac{100}{3.14} \approx 31.85$ feet across

If the bundle has 200 feet of fencing,

$d = \frac{200}{3.14} \approx 63.69$ feet across

If you know the dimensions of the lot where you're putting your garden, you can determine how much fencing to buy and not waste any.

Looking at Norman windows

A Norman window is a rectangle with a half-circle on the top — the third shape in Figure 17-1. It allows more light in at the top and is pleasing to look at. The perimeter of the Norman window is more of a put-together formula. This window is just one example of how you can take pieces of figures and determine a perimeter by adding up the measures of the pieces.

The perimeter of a Norman window is equal to one half the circumference of a circle whose diameter is equal to the width — the distance across the bottom of the window — plus the lengths of the bottom and two sides. $P = \frac{1}{2}\pi w + w + 2s$. The variable w is the width of the bottom and diameter of the half-circle on the top. The s is the length of the sides.

✔ Find the perimeter of a Norman window that has sides of 3 feet and a bottom, or width, of 4 feet. You need this measure to buy caulking material to seal out drafts.

$P = \frac{1}{2}\pi w + w + 2s = \frac{1}{2}(\pi \cdot 4) + 4 + 2(3)$

$= 2\pi + 4 + 6 = 2\pi + 10 \approx 16.28$ feet

You may have to do more calculations in the hardware store, if the tube of caulk says it covers 10 yards.

Spreading Out: Area Formulas

Area is a measure of how many two-dimensional units (squares) a particular object or surface covers — how much flat space it occupies. Usually, area is given in square inches, square centimeters, square feet, or square miles, and so on.

Picture a floor covered with square tiles. If each tile is 1 foot by 1 foot, counting the number of tiles tells you how large the floor is in square feet. In the real world, most floors aren't covered with tiles that are a convenient one-foot square. Most tile floors have pieces of tiles on the edges and in the corners and around things that are strange shapes. So area formulas, such as the ones in this section, help you do the figuring.

In the previous section, the perimeter formulas deal with linear measure. Linear measure is just one dimension. It's from one place to another. There's no breadth to it. You measure it with a ruler or yardstick or tape measure. Square measurements are used to measure area. They take two measures — one along a side and a second perpendicular (90 degrees) to that side.

Laying out rectangles and squares

Rectangles and squares have basically the same area formulas because they both have square corners and the same lengths on opposite sides. The general procedure here is just to multiply the measure of the length times the measure of the width. The product of two sides that are next to one another is the area.

Finding the area of a rectangle

Most rooms in homes and offices are rectangular in shape. Desks and tables and rugs are usually rectangular, also. This makes it easy to fit furniture and other objects in the room.

The area of a rectangle is its length times its width:

$$A = l \cdot w$$

The area of a rectangle 10 feet long and 6 feet wide is $10 \cdot 6 = 60$ square feet.

✔ A garden 85 feet long by 35 feet wide needs some fertilizer. If a bag of fertilizer covers 6 square yards, how much fertilizer do you need?

Note that the measures are different. The garden is measured in feet and the fertilizer coverage is in square yards. Determine how many square feet the garden is. Then convert the fertilizer coverage to square feet per bag.

$$A = l \cdot w = 85 \cdot 35 = 2{,}975 \text{ square feet}$$

Now, how many square feet are there in a square yard? If a yard is equal to 3 feet, then a square yard is 3 feet by 3 feet, so the area is $3^2 = 9$ square feet. There are 9 square feet in a square yard. A bag of fertilizer covers 6 square yards, so that's $6 \cdot 9 = 54$ square feet per bag.

Divide the 2,975 square feet by 54 square feet.

$$\frac{2,975}{54} = 55\frac{5}{54}$$

You can buy 56 bags and have a little left over or buy 55 bags and skimp a little in some places.

Filling in a square

The area of a square is the length of its side squared, so the formula is:

$$A = s^2$$

The area of a square is simple to find. The area of a square is a perfect square. How much nicer can this get? A square that is 8 inches on a side has an area of $8^2 = 64$ square inches. A square $\frac{1}{2}$ a foot long on a side has an area of $\left(\frac{1}{2}\right)^2 = \frac{1}{4}$ a square foot.

You can use the formula for the area of a rectangle, $A = s \cdot s$. But because the length and width are the same measure in a square, the variables are the same. So, it's easier to say $A = s^2$.

Tuning in triangles

Finding the area of a triangle can be a bit of a challenge. Basically, a triangle's area is half that of an imaginary rectangle that the triangle is part of. However, it isn't always easy or necessary to find the length and width of this hypothetical rectangle — you just need a measurement or two from the triangle.

The traditional formula for finding the area of a triangle involves the length of the *base,* or bottom, and the *height,* the perpendicular distance from the base. Finding the area of a triangle is easy if you can use a ruler to find the height, but that isn't always practical. So, you have another option — Heron's formula, covered later in this section.

Going the traditional route

The area of a triangle is equal to half the product of the measure of the base of the triangle times the height of the triangle. $A = \frac{1}{2}bh$

The base is the length of the bottom that the height is drawn down to. The height is the length from the top angle down perpendicular to the base. It forms a right angle (90 degrees) with the base. $A = \frac{1}{2}b \cdot h$

You use this rule when it's possible to make these measurements — when you can draw the height perpendicular to the base and measure both of them.

✏ Find the area of a triangle 20 feet long with a height of 13 feet.

$$A = \frac{1}{2}(20)(13) = \frac{1}{2}(260) = 130 \text{ square feet}$$

✏ Find the area of a right triangle that has sides of 5 inches, 12 inches, and 13 inches.

A right triangle has two sides perpendicular to one another (the two shorter sides). So you can use one side as the base and the other side as the height. The longest side in a right triangle is the hypotenuse, and it's never part of the right angle. So the two measures that you want are the 5 inches and 12 inches.

$$A = \frac{1}{2}(5)(12) = \frac{1}{2}(60) = 30 \text{ square inches}$$

Soaring with Heron's formula

The area of a triangle is equal to the square root of the product of four values: the semi-perimeter (half the perimeter), the semi-perimeter minus the first side, the semi-perimeter minus the second side, and the semi-perimeter minus the third side. Let s represent the semi-perimeter and a, b, and c represent the measures of the sides. $A = \sqrt{s(s-a)(s-b)(s-c)}$

When you're trying to find the area of a huge triangle — say a big park — or if you can't measure any angles to draw a line perpendicular to one of the sides, then you can find the area simply by measuring the three sides and using Heron's formula: $A = \sqrt{s(s-a)(s-b)(s-c)}$

To use this formula, you need the measures of the three sides, a, b, c and the length of the semi-perimeter, s. The *semi-perimeter* is half the perimeter. And, remember, the perimeter is the distance around the triangle — the sum of the measure of the three sides.

✏ Find the area of a triangle with sides of 7 inches, 24 inches, and 25 inches.

Let $a = 7$, $b = 24$, $c = 25$. The perimeter is $P = 7 + 24 + 25 = 56$ inches, so the semi-perimeter, $s = \frac{56}{2} = 28$ inches.

$$A = \sqrt{s(s-a)(s-b)(s-c)} = \sqrt{28(28-7)(28-24)(28-25)} =$$
$$\sqrt{28(21)(4)(3)} = \sqrt{7056} = 84 \text{ square inches}$$

✏ Find the area of a triangle with sides 2, 3, and 4 feet.

Not all answers are going to come out as nicely as the one in the previous example. I purposely picked a right triangle so you'd get an answer that's a nice integer. This example is more typical.

If the sides are 2, 3, and 4, then $a = 2$, $b = 3$, $c = 4$ and $s = \frac{1}{2}(2 + 3 + 4) = 4.5$. So

$$A = \sqrt{s(s-a)(s-b)(s-c)} = \sqrt{4.5(4.5-2)(4.5-3)(4.5-4)} =$$
$$\sqrt{4.5(2.5)(1.5)(.5)} = \sqrt{8.4375} \approx 2.905 \text{ square feet}$$

Going around in circles

The area of a circle is tied to both the radius of the circle and the value of pi.

The formula for the area of a circle is pi (about 3.14) times the radius squared:

$$A = \pi \cdot r^2$$

✔ Find the area of a circular disk that is 50 feet across.

First you need to find the radius. If the circle is 50 feet across, that's the measure of the diameter, all the way across. So the radius is half that or 25 feet.

$$A = \pi \cdot r^2 = \pi \cdot 25^2 = (3.14) \cdot 625 = 1,962.5 \text{ square feet}$$

Why are there 360 degrees in a circle (and not 100 or 1,000)?

The measure used for angles, the degree, is entirely an invented unit. It didn't occur naturally like some other ratios, such as pi. Anything could have been used: any size and any name.

It appears that the degree had its origins in ancient Babylon and is derived from their base-60 number system. (Base-60 has sixty different digits as opposed to our ten — how would you like to learn that one?) The base-60 system has had its influence. Hours and minutes are divided into 60. There are 60 minutes in a degree (that's how degrees are subdivided).

And, really, 360 is a nice total number of degrees to use. It divides the circle into more equal parts than would be possible if there were 100 or 1,000 degrees. The number 360 divides evenly by 2, 3, 4, 5, 6, 8, 9, 10, 12, 15, 18, 20, 24, 30, 36, 40, 45, 60, 72, 90, 120, and 180 — being divisible by 3 really makes a difference. The number 100 divides evenly only by 2, 4, 5, 10, 20, 25, and 50.

Pumping Up with Volume Formulas

Area is a two-dimensional figure or representation. It's a flat region. Volume is three-dimensional. Unlike your last loser boyfriend or girlfriend, it has depth. To find volume, you measure across, front-to-back, and up-and-down.

With volume, you count how many cubes (picture sugar cubes) you can fit into an object. These cubes can be one inch on each edge, one centimeter on each edge, one foot on each edge, or however big they need to be. And, in keeping with the cube theme, you measure volume in cubic inches, cubic feet, cubic centimeters, and cubic whatevers.

Prying into prisms and boxes

The volume of a *prism,* better known as a *box,* is one of the simplest to find in the world of volume problems. The bottom and top of a prism have exactly the same measurements. The distance from the top to bottom is the same, no matter where you measure, as long as you keep that distance perpendicular to both top and bottom.

The formula for finding volume — $V = l \cdot w \cdot h$ — means that the Volume is equal to the product of the length, *l*, times the width, *w*, times the height, *h*.

✔ Find the volume of a box that is 4 feet long, 3 feet wide, and 9 feet high.

   $V = l \cdot w \cdot h = 4 \cdot 3 \cdot 9 = 108$ cubic feet

   That's 108 cubes, all 1 foot by 1 foot by 1 foot, that fit into the box.

✔ If you're buying a 12-cubic-foot refrigerator, what are the dimensions (how big is it)?

   There are an infinite number of ways to multiply three numbers together to get 12. Go through some integers and some fractions. Try to picture what the refrigerator would look like with these dimensions.

   • $12 = 1 \cdot 1 \cdot 12$ (1 foot long, 1 foot wide and 12 feet tall!)

   • $12 = 2 \cdot 1 \cdot 6$ (2 feet long, 1 foot wide and 6 feet tall)

   • $12 = 2 \cdot 3 \cdot 2$ (2 feet long, 3 feet wide and 2 feet tall)

   • $12 = 2 \cdot (1.5) \cdot 4$ (2 feet long, 1.5 feet wide, and 4 feet tall)

   • $12 = 1.5 \cdot (1.5) \cdot 5\frac{1}{3}$ (1.5 feet long and wide and $5\frac{1}{3}$ feet tall)

   Which refrigerator would you want? How tall are you? How far can you reach into the back?

Cycling cylinders

Cylinders were my brother's favorite shape when he was in the Navy on an aircraft carrier in the middle of the ocean. Being the wonderful sister that I am, I would send him chocolate chip cookies that fit exactly into a three-pound coffee can. Imagine a stack of chocolate chip cookies coming to you every couple of weeks. Was he ever popular on that ship!

The formula for the volume of a cylinder is $V = \pi r^2 h$. The volume is equal to pi times the radius (halfway across a circle) squared times the height.

A cylinder is a solid figure with a circle for a base. A can of tuna, an oil storage tank, a can of peas, a roll of toilet paper, and, of course, a coffee can are all examples of cylinders. The top and bottom are circles, and the height of a cylinder is the distance between the circles.

The formula for finding the volume of a cylinder is much like that for a prism because you multiply the area of the base times the height.

- **Prism:** $V = l \cdot w \cdot h$

   The formula to find the area of a base is $l \cdot w$; height is h.
- **Cylinder:** $V = \pi \cdot r^2 \cdot h$

   The formula to find the area of a circular base is $\pi \cdot r^2$; height is h.

To find the volume of a cylinder, you need the radius of the top and bottom, and you need the height.

This formula tells you how many cubes will fit in the cylinder — like putting square pegs in a round hole — just trim them a bit.

The volume of a cylinder is usually given in terms of liquid measures, such as quarts, gallons, or fluid ounces, but it can be given in terms of cubes. One fluid ounce is about 1.8 cubic inches. The volume would be how many cubes fit inside the figure. I'm going to give an example in cubic feet, one foot on each side, in this case.

- You want to figure out how high your aboveground swimming pool will have to be if it can have a radius of 10 feet and it's to hold so many cubic feet of water.

   Solving the formula for h, the height:

   $$V = \pi r^2 h$$

   Divide each side by πr^2

   $$\frac{V}{\pi r^2} = \frac{\pi r^2 h}{\pi r^2}$$

$$\frac{V}{\pi r^2} = h$$

The height is equal to the volume divided by the product of π and the radius squared.

Put in the 10 feet for r

$$h = \frac{V}{100\pi}$$

If your pool is going to hold 400 cubic feet of water, then

$h = \dfrac{400}{100\pi} = \dfrac{4}{\pi} \approx \dfrac{4}{3.14} \approx 1.27$ feet deep. You can go wading!

If your pool is going to hold 800 cubic feet of water, then

$h = \dfrac{800}{100\pi} = \dfrac{8}{\pi} \approx \dfrac{8}{3.14} \approx 2.55$ feet deep.

If your pool is going to hold 1,000 cubic feet of water, then

$h = \dfrac{1000}{100\pi} = \dfrac{10}{\pi} \approx \dfrac{10}{3.14} \approx 3.18$ feet deep.

If your pool is going to hold 2,000 cubic feet of water, then

$h = \dfrac{2000}{100\pi} = \dfrac{20}{\pi} \approx \dfrac{20}{3.14} \approx 6.37$ feet deep.

It takes a lot of water to raise the level in this tank.

Scaling a pyramid

A pyramid is an easy thing to describe because everyone has a mental picture of what a pyramid looks like. Technically, a *pyramid* is an object with a base (bottom) and triangles coming up from each side of the base to meet at a point.

The pyramids in Egypt have a square for the bottom and same-sized triangles on the sides — at least, that's how they started. The wind and sand have eroded the tops so the Egyptian pyramids don't come to a point anymore. But, the base of a pyramid can be an equilateral triangle (all sides are the same length), a square, a regular pentagon (five sides, all the same length), and so on. The examples in this chapter, however, stick with square bases.

The volume of a pyramid is one-third the area of the base multiplied by the height:

$$V = \frac{1}{3}(\text{area of base}) \cdot h$$

The volume of a pyramid isn't too hard to determine. So, it pays to be up on the formulas for finding area. Check out the "Spreading Out: Area Formulas" section earlier in this chapter for refreshers, though the area formulas for squares and triangles are included in the following formulas:

✔ **Square base:** $V = \frac{1}{3} s^2 \cdot h$

✔ **Equilateral triangular base:** $V = \frac{1}{3} \times \frac{\sqrt{3}\, s^2 h}{4} = \frac{\sqrt{3}\, s^2 h}{12}$

This first example is a sort of historical challenge:

✔ Find the original volume of the Great Pyramid. The Great Pyramid *originally* had a square base with each side measuring 756 feet and a height of 480 feet.

$$V = \frac{1}{3} s^2 \cdot h = \frac{1}{3} (756)^2 \cdot 480 = 91{,}445{,}760 \text{ cubic feet}$$

Perhaps you can more easily relate to finding the volume of a pyramid-shaped tent whose square sides are each 6 feet long and whose height is 10 feet.

$$V = \frac{1}{3} s^2 \cdot h = \frac{1}{3} (6)^2 \cdot 10 = 120 \text{ cubic feet.}$$

Even that's pretty big! A cubic foot is the size of a small packing box — and there would be 120 of them!

Pointing to cones

The formula for the volume of a cone is one-third times pi times the radius squared times the height:

$$V = \frac{1}{3} \pi \cdot r^2 \cdot h$$

The formula for finding the volume of a cone should look familiar for two reasons. First, it has the $\frac{1}{3}$, like the pyramid formula has. The one-third factor is common when a figure goes up into a single point. The other familiar part is the $\pi r^2 h$, which is the formula for finding the volume of a cylinder. You can think of a cone as being just a cylinder that was whittled away. The pointy-bottomed ice-cream cone is a classic cone shape, as are traffic pylons.

The two examples below should be situations you recognize from personal experience.

✔ What is the volume of a cone-shaped tent that has a diameter of 18 feet and a height of 20 feet?

If the diameter is 18 feet, then the radius is 9 feet.

$$V = \frac{1}{3}\pi \cdot r^2 \cdot h = \frac{1}{3} \pi(9)^2 \cdot 20 = 540\pi \approx 1{,}696 \text{ cubic feet}$$

✔ How much ice cream can you fit into a sugar cone that has a diameter of 3 inches and is 6 inches tall? (This doesn't count the ice cream that you can put on top of the filled cone. That comes next.)

$$V = \frac{1}{3} \pi \cdot r^2 \cdot h = \frac{1}{3} \pi(1.5)^2 \cdot 6 = 4.5\pi \approx 14 \text{ cubic inches of ice cream inside}$$

Rolling along with spheres

The formula to determine the volume of a sphere is four-thirds times pi times the radius cubed:

$$V = \frac{4}{3}\pi \cdot r^3$$

A sphere is a familiar shape. Basketballs and baseballs, marbles and globes are all spheres. As you can see from the formula, you need only one thing to find the volume of a sphere: the radius, which is the distance from the center of the sphere to the outside.

Finding the volume of a sphere can be helpful when buying a tank of helium or a container of ice cream.

- ✔ What is the volume of a ball that has a diameter of 18 inches?

  A diameter of 18 inches means that the radius of the ball is 9 inches.

  $$V = \frac{4}{3}\pi \cdot r^3 = \frac{4}{3}\pi \cdot 9^3 = 972\pi \approx 3,052 \text{ cubic inches}$$

  Back to that ice-cream cone. If the cone's opening is 3 inches, then you can certainly fit a sphere (scoop) of ice cream on it that has a diameter of 4 inches. It's okay if it's bigger — as long as you lick fast.

- ✔ What is the volume of a sphere with a diameter of 4 inches?

  $$V = \frac{4}{3}\pi \cdot r^3 = \frac{4}{3}\pi \cdot 2^3 = 10\frac{2}{3}\pi \approx 33.5 \text{ cubic inches}$$

Going the Distance with Distance Formulas

You've been on a slow boat to China for a couple of days and want to know how far you've come. Or you want to figure out how long it will take to drive to grandmother's house for dinner. Or maybe you want to know how fast a train travels if it gets you from Toronto, Ontario, to Miami, Florida, in 18 hours. The distance-equals-rate-multiplied-by-time formula can help you find the answer to all these questions.

Figuring out how far and how fast

The formula $d = r \cdot t$ means the **distance** traveled is equal to the **rate** of speed times how long it takes (the **time**).

To change the formula to one that you can use to find out how long it will take to get somewhere (say, to grandma's house), you solve for time, *t*, by dividing each side of the equation by *r,* the rate. Now the formula reads that the time it takes to make the trip is equal to the total distance traveled divided by the rate of speed.

You can plug in the distance to grandma's and the speed you'll be traveling to find out how long it will take you.

It's decision time. What's the speed limit? How's the traffic?

Similarly, if you want to know how fast an express train travels the 1,492 miles from Toronto to Miami in just 18 hours, you solve the distance-rate-time formula for rate, by dividing each side of the equation by *t,* or time.

Now you can divide the distance (number of miles) by the time (18 hours) to find the rate. That's one fast train! (Even if it's only in my head.)

The simple formula allows you to determine the distance traveled or how long it took to go a distance or how fast you traveled. The only caution is with the *rate* part. If you say that you drove 300 miles in 6 hours, you could use the formula and say that $300 = r \cdot 6$, $r = 50$, or that the rate you traveled was 50 miles per hour. In real life, though, you know that you didn't drive 50 miles per hour the *whole time*. There was the start up at 0 miles per hour and the slow down to stop. And, certainly, you stopped along the way for refreshments! Realistically, the rate is an average speed.

The following problems use the distance formula, but solve for rate and time, not distance, just to show how versatile one little formula can be.

✔ What is the average speed of an airplane that can go 2,000 miles in 4.8 hours?

$d = r \cdot t$

$2,000 = r \cdot 4.8$

$r = \dfrac{2,000}{4.8} = 416\frac{2}{3}$ miles per hour

✔ How long did it take the settlers to get from St. Louis to Sacramento if they could average 12 miles per day? The distance between the two cities is about 1,980 miles.

$d = r \cdot t$

$1,980 = 12 \cdot t$

$t = \dfrac{1,980}{12} = 165$ days

That's almost half a year. You can drive it now in less than 40 hours.

Sailing along in high seas

You're watching an old John Wayne movie, with the Duke playing a ship's captain. (Do *not* tell me you don't know who John Wayne is. He won the war — several of them, several times.) Anyway, in the movie, the boat can go only so many knots. (Have you ever wondered what in the world a knot is doing in this movie?)

Miles are measured differently in the water than they are on land. There are two relations you can use as formulas.

> 1 nautical mile = 1.51 statute miles
>
> 1 knot = 1.152 miles per hour

The first relation is just one measure changed to another measure. A *statute* mile is just a normal mile — the one that's 5,280 feet. The word *statute* is used when it's being compared to another type of mile, like the *nautical* mile.

After looking at these next examples, perhaps next time you watch a movie that's about sailing the high seas, you'll understand what they're talking about with some of the nautical lingo.

- How far has a bird flown if it's covered 16 nautical miles?

    Write a proportion and cross-multiply.

    $$\frac{1 \text{ nautical mile}}{16 \text{ nautical miles}} = \frac{1.51 \text{ statute miles}}{x \text{ statute miles}}$$

    x statute miles = 16 · 1.51 = 24.16 statute miles

- How fast is John Wayne's ship going if it's going 14 knots?

    This is more complicated. It changes a single value, *knot,* into a rate of speed, miles per hour.

    $$\frac{1 \text{ knot}}{14 \text{ knots}} = \frac{1.152 \text{ miles per hour}}{x \text{ miles per hour}}$$

    x miles per hour = 14 · (1.152) = 16.128 miles per hour.

(Perhaps you'll just remember that the speed in miles per hour is just "a little more than" the number of knots.) For more distance problems, see Chapter 18.

Calculating Interest and Percent

Percentages are a part of our modern vocabulary. You probably hear or say one of these phrases every day:

- ✔ The chance of rain is 40%.
- ✔ There was a 2% rise in the Dow Jones Industrial Average.
- ✔ The grade on your test is 99%.
- ✔ Your height puts you in the 80th percentile.

Percent is used to express fractions as equivalent fractions with a denominator of 100. The percent is what comes from the numerator of the fraction — how many out of 100.

- ✔ $80\% = \dfrac{80}{100} = 0.80$
- ✔ $16\dfrac{1}{2}\% = \dfrac{16.5}{100} = 0.165$
- ✔ $2\% = \dfrac{2}{100} = 0.02$

You use percents and percentages in the formulas that follow. Change the percentages to decimals so that they're easier to multiply and divide. To make this change, you move the decimal point in the percent two places to the left. If no decimal point is showing, assume it's to the right of the number.

Compounding interest formulas

Figuring out how much interest you have to pay, or how much you're earning, is simple with the formulas in this section.

Figuring simple interest

Simple interest is used to determine the amount of money earned in interest that is not compounded. It's also used to figure the total amount to pay back when buying something on time. Simple interest is basically a percentage of the total amount. It's figured on the beginning amount only — not on any changing total amount that can occur as an investment grows. To take advantage of the growth in an account, use compound interest.

The amount of simple interest earned is equal to the amount of the principal *(P)* (starting amount) times the rate of interest *(r)*, which is a percent, times the amount of time *(t)* involved.

The formula to calculate simple interest is:

$$I = Prt$$

✔ What is the amount of simple interest on $10,000 when the interest rate is $2\frac{1}{2}$% and the time period is $3\frac{1}{2}$ years?

$$I = Prt$$

$$I = 10,000 \cdot (0.025) \cdot 3.5 = 875$$

The interest is $875.

✔ You're going to buy a television "on time." The appliance store will charge you 12% simple interest. You add this onto the price of the television and pay back this whole amount in "24 easy monthly payments." Twenty-four months = 2 years; so $t = 2$. The television costs $600. How much are your "easy payments"?

$$I = 600 \cdot (0.12) \cdot 2 = 144$$

The interest is $144. Add that onto the cost of the television and the total is $744. Divide this by 24, and the payments are $\frac{744}{24} = 31$; that's $31 per payment. Such a deal.

Tallying compound interest

Compound interest is used when determining how much you have in your savings account after a certain amount of time. *Compound interest* is called compound because the interest earned is added to the beginning amount before the new interest is figured. The amount of times per year the interest is *compounded* depends on your account, but many savings accounts compound *quarterly,* or four times per year.

By leaving the earned interest in your account, you're actually earning more money because the interest is figured on the new, bigger sum.

The formula for compound interest is:

$$A = P\left(1 + \frac{r}{n}\right)^{nt}$$

The total amount in the account is equal to the principal (starting amount) times the sum of one plus the percentage rate divided by number of times it's compounded, all raised to a power equal to the product of the number of compounding times per year and the number of years. (Whew!)

✔ How much is there in an account that started with $5,000 that has been earning interest for the last 14 years at the rate of 6%, compounded quarterly?

The *principal* is $5,000; the *rate* is 6%, or 0.06, the *number* of times per year it's compounded is 4; the *time* in years is 14.

$$A = 5,000\left(1 + \frac{.06}{4}\right)^{4 \cdot 14}$$

Carefully work from the inside out. Divide the 0.06 by 4 and add it to the 1. At the same time, multiply the 4 and 14 in the exponent to make it simpler.

$$A = 5,000(1.015)^{56}$$

By the *order of operations,* raise to the power first.

$$A = 5,000(2.30196) = 11,509.82$$

The amount of money more than doubled. Compare this to the same amount of money earning simple interest. Using the simple interest formula, $I = Prt = 5,000(0.06)(14) = \$4,200$. Add this interest onto the original $5,000, and the total is $9,200. That's more than $2,000 less than the amount earned using compound interest.

Look at an even more dramatic example of the power of compounding:

✔ Suppose that you get a letter from the Bank of the West Indies. They claim that some ancestor of yours came over with Columbus, deposited one dollar with them, and then was lost at sea on the way home. His dollar has been sitting in their bank, earning interest at the rate of $3\frac{1}{2}$% compounded quarterly. This is becoming a nuisance account because fees have to be collected; they want to charge this account the current fee rate of $25 per year — retroactively. Do you want to claim this account?

At first, you may say, "No way! I'd owe money." Then you get out your trusty calculator and figure this out. If your ancestor came over with Columbus in 1492, and if you got the letter in the year 2002, what exactly are you looking at?

The principal is $1; the interest rate is $3\frac{1}{2}$% compounded 4 times per year. This money has been deposited for 510 years, but that means 510 years of $25 service charges.

$$A = 1\left(1 + \frac{.035}{4}\right)^{4 \cdot 510} = 1(1.00875)^{2,040} = 52,292,254.90$$

That's over 52 million dollars for an initial deposit of $1.

Subtracting the service charges: $25 \cdot 510 = 12,750$; paying $13,000 is minor. Take the money!

Gauging taxes and discounts

You can figure both the tax charged on an item you're buying and the discount price of sale items with percentages.

✔ **Total price** = cost of item $\times$ (1 + percent tax as a decimal)

✔ **Discounted price** = original cost $\times$ (1 – percent discount as decimal)

✔ **Original price** = discount price $\div$ (1 – percent discount as decimal)

FUN FACT

Probability and birthdays

A really interesting occurrence can be confirmed with probability. If you're in a room containing 24 people, there's better than a 50% chance that two of the people in the room have the same birthday — the same month and day. Increase the number of people in the room, and the chances are even greater for a birthday match. Just think of it — 365 or 366 birthdays to choose from and it takes only 24 people to make a better-than-even bet that two have the same birthday.

The mathematical branch of probability was born of a gambling question. A friend of the 17th-century French mathematician Blaise Pascal was a professional gambler and wrote to Pascal asking why a particular roll of the die happened so often and also how to split the pot of an interrupted dice game. Pascal and mathematician Pierre de Fermat wrote letters back and forth and subsequently founded the theory of probability.

All consumers are faced with taxes on purchases and hope to find situations where they can buy things on sale. It pays to be a wise consumer. Look at these examples.

- The $24,000 car you want is being discounted by 8%. How much will it cost now with the discount? Be sure to add the 5% sales tax.

  Discounted price = $24,000 \times (1 - 0.08) = 24,000 \times 0.92 = \$22,080$

  Total price = cost of item $\times$ (1 + percent tax as a decimal)

  Total price = $22,080 \times (1 + 0.05) = \$23,184$

- The shoes you're looking at were discounted by 40% and then that price was discounted another 15%. What did they cost, originally, if you can buy them for $68 now?

  If the price now was discounted 15%, find the amount they were discounted from first (the first discount price).

  $$\text{Original price} = \frac{\text{discount price}}{1 - \text{percent discount as decimal}}$$

  $$\text{"First discounted price"} = \frac{68}{(1 - 0.15)} = \frac{68}{.85} = \$80$$

  $$\text{Original price} = \frac{80}{(1 - 0.40)} = \frac{80}{.60} = \$133.33$$

  The discount of 40% followed by 15% is not the same as a discount of 55%. A 55% discount would have resulted in $60 shoes.

Working Out the Combinations and Permutations

Combinations and permutations are methods and formulas for counting things. You may think that you have that "counting stuff" mastered already, but do you really want to count the number of ways in the following?

✔ How many different vacations can you take if you plan to go to three different states on your next trip?

✔ How many different ways can you rearrange the letters in the word *smart* — and how many of them actually make words?

✔ How many different ways can you pick 6 numbers out of 54, and can you put $1 on each way to win the lottery?

You could start making lists of the different ways to accomplish the preceding problems, but you'd quickly get overwhelmed and perhaps a little bored. Algebra comes to the rescue with some counting formulas called combinations and permutations.

Counting down to factorials

The main operation in combinations and permutations is the *factorial* operation. This is really a neat operation. It only takes one number to perform it. The symbol that tells you to perform the operation is an exclamation mark: "!". When I write, "6!", I don't mean, "Six, wow!" Well, I suppose I might say that if my dog had six puppies. But, in a math context, the exclamation mark has a specific meaning.

$$6! = 6 \cdot 5 \cdot 4 \cdot 3 \cdot 2 \cdot 1 = 720$$

$$4! = 4 \cdot 3 \cdot 2 \cdot 1 = 24$$

$$7! = 7 \cdot 6 \cdot 5 \cdot 4 \cdot 3 \cdot 2 \cdot 1 = 5,040$$

The *factorial* of any whole number is the number you get by multiplying the whole number by every counting number smaller than it. Remember, the counting numbers are 1, 2, 3, 4, and so on.

$$n! = n \cdot (n - 1) \, (n - 2) \, (n - 3) \ldots$$

Factorial works when *n* is a whole number; that means that you can use numbers such as 0, 1, 2, 3, 4, . . . One surprise, though, is the value of 0!. Try it on a calculator. You get 0! = 1.

Googol

A googol is a very big number. It was actually named a googol by a nine-year-old who was asked to make up a name for a very big number. A googol is a 1 followed by 100 zeros. It's 10^{100}. This number is supposed to be bigger than the number of electrons in the entire universe. A googol is a bit bigger than 69!. (That's 69 factorial, not excitement.) If you have a scientific calculator, try entering 69! and then 70!. Most calculators can handle the 69! but give you an "overflow" or "error" message on the 70! They just can't handle a googol. A googolplex is an even bigger number. It's 10 raised to the googol power, 100^{googol}.

Counting on combinations

Combinations tell you how many different ways you can choose *some* from a group — you can choose anywhere from one to all the group. You can

- ✔ Figure out how many ways to choose three states to visit.
- ✔ Figure out how many ways there are to choose 6 numbers out of a possible 54 numbers.
- ✔ Figure out how to choose 8 astronauts out of a group of 40 candidates.

Combinations don't tell you *what* is in each of these ways, but they tell you how many ways there are. After you find them all, you can stop trying different arrangements.

The number of *Combinations* of r things taken from a total possible of n things is

$$_nC_r = \frac{n!}{r!(n-r)!}$$

The subscripts on the *C* tell two things. To the left, the n indicates how many things are available altogether. The subscript to the right, the r, tells how many will be chosen from all those available. The computation involves finding n factorial divided by the product of r factorial times the difference of n and r factorial.

These examples show you how to solve counting questions using the formula for the number of combinations.

✔ Find the number of different ways to choose 3 states out of 50.

The total number of things, n, is 50. The number you want to choose out of the 50 is r, or 3.

$$_{50}C_3 = \frac{50!}{3!(50-3)!}$$

You need a calculator for this one, but

$$_{50}C_3 = \frac{50!}{3!(50-3)!} = \frac{50!}{3!47!} = 19,600$$

These are 19,600 different vacations if you choose three states. I'll start listing them.

Alabama, Alaska and Arizona; Alabama, Alaska and Arkansas; Alabama, . . . Okay, enough.

It doesn't take long to see what a task this is. And this doesn't even take into account the *order* that the states are visited in. That would be six times as many ways — and that's *permutations*.

✔ How many ways are there to select 6 numbers out of a possible 54?

$$_{54}C_6 = \frac{54!}{6!(54-6)!} = \frac{54!}{6!48!} = 25,827,165$$

Guess that's too many to buy a ticket for each combination in a lottery game — even if the machines could print them all out.

✔ How many ways are there to select 8 astronauts out of 40?

$$_{40}C_8 = \frac{40!}{8!(40-8)!} = \frac{40!}{8!32!} = 76,904,685$$

These numbers are all pretty big. How about some examples where they're more reasonable?

✔ How many ways are there to choose two books from a shelf where there are 7 books?

$$_7C_2 = \frac{7!}{2!(7-2)!} = \frac{7!}{2!5!} = \frac{5,040}{240} = 21$$

Okay, that's more like it. You can choose *Tom Sawyer* and *Tale of Two Cities,* or *Tom Sawyer* and *Atlas Shrugged,* and so on.

Ordering up permutations

Permutations are like combinations. The main difference is that in permutations the *order* matters. If you choose a vacation that involves trips to Alaska, Arizona, and Alabama, there are six different ways to arrange the visits.

Alaska, Arizona, Alabama	Alaska, Alabama, Arizona
Arizona, Alaska, Alabama	Arizona, Alabama, Alaska
Alabama, Arizona, Alaska	Alabama, Alaska, Arizona

Just like with combinations, finding the *number* of permutations doesn't tell you *what* they are, but it does tell you when you can finish with your list.

The number of permutations of *r* things taken from a total possible of *n* things is

$$_nP_r = \frac{n!}{(n-r)!}$$

The subscripts on the *P* tell two things. To the left, the *n* indicates how many things are available altogether. The subscript to the right, the *r*, tells how many will be chosen from all those available. The computation involves finding *n* factorial divided by the difference of *n* and *r* factorial.

Notice that the only difference between this formula and the one for combinations is that the *r*! is missing from the denominator. This makes the denominator a smaller number, which makes the end result a bigger number. When things are put in specific orders, there are more ways to do it.

Here are some examples using the counting method called permutations.

- How many ways are there to choose 2 books out of 7 on the shelf, if the order that you select them matters (which first and which second)?

  $$_7P_2 = \frac{7!}{(7-2)!} = \frac{7!}{5!} = 42 \text{ ways to choose the books}$$

- How many different arrangements are there of the letters in the word "smart"? How many of these are actually words themselves?

  There are 5 letters in the word, and all 5 will be used each time.

  $$_5P_5 = \frac{5!}{(5-5)!} = \frac{5!}{0!} = \frac{120}{1} = 120 \text{ different arrangements}$$

I don't think I want to write all of them, but there *are* some words that can be formed from SMART: TRAMS, MARTS, . . . That may be it.

Formulating Your Own Formulas

A formula is just an equation or relationship that is accepted as true and used over and over in common applications. This chapter deals with many of the most frequently used formulas. But there's no end to the number of formulas in the world. You can make up and use your own formula.

The basics to writing your own formula are:

1. Assign letters to the variables representing numbers in your situation.

2. Write expressions using operations and comparisons of the variables.

3. Try the formula for accuracy. Then use it again and again.

Here's an example of creating a formula that's never been used before (at least not by me).

You're planning a wedding reception and the banquet hall has rectangular tables that seat eight people, round tables that seat six people, and square tables that seat four people. You want a formula for determining how many people can be seated using all these shapes of tables.

1. **Assign letters.**

   Let P represent the total number of people that can be seated; let r represent the number of rectangular tables; let c represent the number of round (circular) tables; let s represent the number of square tables.

2. **Write an expression.**

   The number of people at a table will have to multiply the number of each kind of table. All those will get added.

   $$P = 8r + 6c + 4s$$

3. **Try it out.**

   If you choose to have five rectangular, ten round, and seven square tables, then you can seat:

   $$P = 8(5) + 6(10) + 4(7) = 40 + 60 + 28 = 128 \text{ people}$$

The different size tables allow you some flexibility when seating some of the more, shall we say, interesting people.

Chapter 18

Sorting Out Story Problems

In This Chapter

▶ Using general suggestions for solving story problems

▶ Mixing it up with mixture story problems

▶ Keeping your distance with distance problems

▶ Using right triangles

▶ Circling the problem

Story problems can be one of the least favorite things for algebra students. Although algebra and its symbols, rules, and processes act as a door to higher mathematics and logical thinking, story problems give you immediate benefits and results in *real-world* terms. I recognize that some story problems seem a bit contrived, which is why I don't include age problems such as: "If Henry is three times as old as George was when George was 5 years older than Beth . . ." Who cares? And, you won't find any consecutive integer problems in this chapter. Consecutive integer problems read something like: "Find three consecutive even integers whose sum is 102." By the way, the answer is: 32, 34, and 36. These types of problems are good for developing the logical patterns necessary for further study in math, but I want to win you over on practicality here, so I leave them out.

Algebra allows you to solve problems (sorry, it won't help with that noisy neighbor) — problems involving how to divide up money equitably or make things fit in a room or divvy up shares. In this chapter, you find some practical applications for algebra.

Setting Up to Solve Story Problems

When solving story problems, the equation you should use or how all the ingredients interact isn't always immediately apparent. Sometimes you have to come up with a game plan to get you started. Just getting started can be a big help. You don't have to use every suggestion in the following list with every problem, but using as many as possible can make the task more manageable.

✔ Draw a picture. It doesn't have to be particularly lovely or artistic. Many folks respond well to visual stimuli, and a picture can act as one.

✔ Assign a variable(s) to represent *how many* or *number of*. You may use more than one variable at first and refine the problem to just one variable later. Remember, a variable can represent only a *number;* it can't stand in for a person, place, or thing.

✔ If you use more than one variable, go back and substitute known relationships for the extra variables. When it comes to solving the equations, you want to solve for just one variable. You can often rewrite all the variables in terms of just one of them. For example, if you let *a* represent the number of Ernie's cookies and *b* represent Bert's cookies, but you know that Ernie has four more cookies than Bert, then *a* can be replaced with *b*+4.

✔ Look at the end of the question or problem. This often gives a big clue as to what is being asked for and what the variables should represent. It can also give a clue as to what formula to use, if a formula is appropriate.

✔ Translate the words into an equation. Replace

- *and, more than, exceeded by* with the plus sign

- *less than, less, subtract from* with the minus sign

- *of, times as much* with the multiplication sign

- *twice* with 2×

- *divided by* with the division sign

- *half as much* with ½ ×

- the verb (*is* and *are,* for example) with the equal sign

✔ Plug in a standard formula, if the problem lends itself to one. Check out the Cheat Sheet at the front of the book for common formulas or go to Chapter 17 for in-depth information.

✔ Draw a picture of containers when doing mixture problems. Label each container with a *quantity* times a *quality*. The *quality* is the strength of the solution or percent interest or price per pound of the item.

After you get set up, you're ready to solve the equation. Make sure to check your answer for *sense* as well as accuracy. Ask yourself whether the answer makes sense in this situation.

Working Around Perimeter, Area, and Volume

Perimeter, area, and volume problems are some of the most practical of all story problems. It's hard to avoid situations in life where you'll be dealing with one or more of these values. Someday, you may want to put up a fence and need to find the perimeter of your yard to help determine how much material you need to buy. Perhaps an addition to the family means adding a room to your home (or building a doghouse). You can use an area formula to figure how much space your new room will take up. Finding a box to contain your present for your Aunt Bea's 80th birthday may require calculating the volume of standard box sizes, and then constructing your own box for that special gift. Lucky for you, standard formulas to deal with all these situations are available, and many of them are in this section.

Parsing the perimeter

Parsing means describing; *perimeter* is the measure around the outside of a region or area. Perimeter is used when you want to put a fence around a yard or some baseboards around a room.

The way to find the perimeter (P) of any shape is to add up the lengths of all the sides. To find the perimeter of a rectangle, add twice the length (l) and twice the width (w). The formula for the perimeter of a rectangle is

$$P = 2l + 2w.$$

To find the area (A) of a rectangle, multiply the length times the width.

$$A = l \cdot w$$

✔ Jim wants to fence in a rectangular part of his yard along the river. He won't need any fence along the side of the yard next to the river, just the other three sides. Jim wants his yard to be twice as long as it is wide, and he'd like it to have an area of 80,000 square feet. What should the dimensions of his yard be, and how much fencing will he need?

The first issue has to do with the area. The formula for the area of a rectangle is $A = l \cdot w$. The area is to be 80,000 square feet, so $80,000 = l \cdot w$.

There are two variables. To change the equation so that it has one variable, go back to the problem where it says Jim wants the length to be twice the width. That means $l = 2w$. Replacing the l with $2w$ in the area formula, you get $80,000 = 2w \cdot w$ or $80,000 = 2w^2$. Solve the equation for w.

First divide by 2: $40,000 = w^2$.

Then take the square root of each side: $w = 200$.

The width is 200. The length is twice that or 400.

If the three sides that need fencing are $200 + 400 + 200$, then the amount of fencing needed is 800 feet.

✔ Janelle has designed a hiking path in the shape of a huge triangle. You start at one corner, walk to the second corner along the first leg (side), walk to the third corner along the second leg, and go back to the first corner along the third leg. The first leg (side) of the path is 3 miles shorter than four times the length of the last leg, and the second leg is 2 miles less than 3 times as long as the last leg. If the entire path is 11 miles long, then how long is each leg?

The perimeter of a triangle is $P = a + b + c$ where the a, b, and c are the lengths of the sides.

Draw a triangle. Label the sides a for the first leg, b for the second leg, and c for the third leg. The total distance of the path is 11 miles, so $a + b + c = 11$. That's too many letters. Because the first two legs are each compared to the length of the last leg, c, you can write the lengths of the first two legs in relation to that last leg:

The first leg, a, is 3 less than four times as long as the last leg. Because the last leg is c, that would make it $4c - 3$.

The second leg, b, is 2 less than 3 times as long as the last leg. That would make it $3c - 2$.

Now you can substitute the equivalents of a and b written with c's into the equation for the length of the path:

$$a + b + c = 11, \text{ so } (4c - 3) + (3c - 2) + c = 11$$

$$8c - 5 = 11$$

$$8c = 16$$

$$c = 2$$

So the last leg is 2 miles, the first leg, $4c - 3$, is $8 - 3 = 5$ miles, and the second leg, $3c - 2$, is $6 - 2 = 4$ miles. Added together, $5 + 4 + 2 = 11$ miles. So, you worked it correctly!

Arranging for area

You may want to buy an area rug. You may meet someone who lives in your area. In both cases, *area* can be interpreted as some measured-off region or surface that has a shape or size. When doing area problems, you can find the area if you know what the shape is, because there are so many nice formulas to use. You just have to match the shape with the formula.

✔ Eli and Esther are thinking of enlarging their family room. Right now, it's a rectangle with an area of 120 square feet. If they increase the length by 4 feet and the width by 5 feet, the new family room will have an area of 240 square feet. What are the dimensions of the family room now, and what will the new dimensions be?

Draw a rectangle, labeling the shorter sides as w and the longer sides as l. The area of a rectangle is $A = l \cdot w$, so, in this case, $120 = l \cdot w$.

The length is going to increase by 4 feet, and the width by 5 feet, so represent the changes as $l + 4$ and $w + 5$.

This new area is 240 square feet, so $240 = (l + 4)(w + 5)$.

In the original room, $120 = l \cdot w$, so you can solve for l and substitute that into the new equation.

$$l = \frac{120}{w}$$
$$\left(\frac{120}{w} + 4\right)(w + 5) = 240$$

Using FOIL (refer to Chapter 10) to simplify the left side,

$$120 + \frac{600}{w} + 4w + 20 = 240$$
$$\frac{600}{w} + 4w = 100$$

To solve this, multiply both sides by w.

$$600 + 4w^2 = 100w$$

Now you have a quadratic equation that can be solved.

$$4w^2 - 100w + 600 = 0$$

Divide through by 4 to make the numbers smaller.

$$w^2 - 25w + 150 = 0$$

The quadratic factors using unFOIL.

$$(w - 15)(w - 10) = 0$$

Now use the multiplication property of zero, where $w - 15 = 0$ or $w - 10 = 0$, to get the solutions:

$$w = 15 \text{ or } w = 10$$

If $w = 15$, then $l = \frac{120}{15} = 8$; the width is increased by 5 and the length by 4, giving you the new dimensions of 20 by 12.

If $w = 10$, then $l = \frac{120}{10} = 12$; the width is increased by 5 and the length by 4 giving you new dimensions of 15 by 16.

Technically, these both work. They both work if you can accept a width that is greater than the length. In the case of $w = 15$, the width is 15 and the length 8. If you're going to hold fast to width being less than length, then only the second solution works: original dimensions of 10 by 12 and new dimensions of 15 by 16.

✔ A playground measures 100 yards by 30 yards. By how much will the area increase if a 6-foot-wide walkway is added all the way around?

Draw a rectangle whose long sides are a little more than three times the length of the shorter sides to represent the 100 yards long and 30 yards wide of the playground. To this rectangle, draw another rectangle around the outside of the first and label the width of the strip between them as 2 *yards*. Why 2 yards? Because 6 feet equals 2 yards and you want to keep the same units and not mix feet and yards.

How much will the area increase?

New area – Old area = Difference

The new area is determined by adding 4 yards to each of the dimensions: $A = (100 + 4)(30 + 4)$. You add 4, because 2 yards was added at each end and above and below.

The old area is determined by multiplying the two dimensions together: $A = (100)(30)$

New area – Old area $= (100 + 4)(30 + 4) - (100)(30) =$ $(104)(34) - 3,000 = 3,536 - 3,000 = 356$ square yards.

The area increased by 356 square yards.

When you figure area, your answer is in *square* yards, feet, meters — the square version of whatever unit of measure you used for length.

Pumping up the volume

An area is a flat measurement. It can be shown on a floor or sports field, in two dimensions. Volume adds a third dimension, as Figure 18-1 shows. Take a room 10 by 12 feet and make the ceiling 8 feet high. You're talking about an area of 10 times 12 or 120 square feet and, with the height, a volume of 10 times 12 times 8 or 960 cubic feet. Volume is measured in cubic measures. The amount of gas in a balloon is a cubic measure. The amount of cement in a sidewalk is a cubic measure.

A cube is a box that has equal length, width and height. Picture a sugar cube, or a set of dice.

Figure 18-1:
Volume is
determined
by
multiplying
length,
width, and
height.

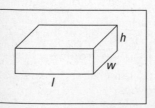

✔ Aunt Sadie got a wonderful deal on some chocolate candies. You're her favorite of all her nieces and nephews, so Aunt Sadie wants to send them all to you. The candies came in a huge plastic bag, but she wants to ship them in a box. The candies take up 900 cubic inches of space. If the box she's going to use to ship them has a 9-by-9-inch bottom, then how high does the box have to be to fit the candy?

The volume of a prism (box) is found by multiplying the length of the box times its width times its height: $V = l \cdot w \cdot h$.

In this case, the bottom is square, and each side of the bottom is nine inches, so, substituting, $V = 9 \cdot 9 \cdot h = 81h$.

Since the candy has a volume of 900 cubic inches, the volume formula becomes $900 = 81h$.

Solving for h, divide each side by 81 to get $\frac{900}{81} = h$, or $11\frac{1}{9}$ inches.

Building a pyramid

Pyramids are one of the more recognizable geometric figures. Young children hear and read about the pyramids of Egypt. You see the pyramid shape in everything from tents to meditation sites to Figure 18-2. If your tent has a pyramid shape, you can find its volume to see if you and your three friends can all fit. You'll want breathing room.

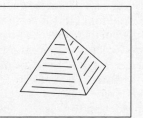

Figure 18-2: Some folks believe pyramids have preservation powers.

The formula for the volume of a pyramid with a square base is $V = \frac{1}{3}x^2 \cdot h$. The x^2 represents the area of the base. In general, the volume of a cube is one-third of the product of the area of the base and the height.

The Great Pyramid of Cheops is a solid mass of limestone blocks. It is estimated to contain 2.3 million blocks of stone. Originally, it had a square base of 756 feet by 756 feet and was 480 feet high, but wind and sand have eroded it over time. Pretend it still has its original dimensions. If each of the blocks is a cube, what are the dimensions of the cubes?

✔ You have only one measure to name — the measure of each edge — so call it x.

First find the volume of the Great Pyramid in cubic feet:

$V = \frac{1}{3} 756^2 \cdot 480 = 91,445,760$ cubic feet.

If each block were a cube 1 foot by 1 foot by 1 foot, there would be over 91 million of them. According to the estimate, there are 2.3 million blocks of stone, so:

$91,445,760 \div 2,300,000 \approx 39.759$

That means that each of the 2.3 million blocks of stone measures more than 39 cubic feet. To find the measure of an edge, which gives you the dimensions, look at the formula for the volume of a cube, $V = s^3$. In this case, assign x to be the length of a side.

So, if $V = 39.759 = x^3$, then $x = \sqrt[3]{39.759} \approx 3.413$.

So each cube would be about 3½ feet on each edge. Picture a huge block of stone longer than a yardstick on each side (some of the stones are reportedly larger than this). Now picture lifting that stone up to the top of the Great Pyramid.

Circling Jupiter

Figuring out how much air you have to expel to blow up a 9-inch balloon involves cubic inches of air, force, propulsion, and all sorts of complicated things, and in the end, do you really care? You just blow until the balloon is full. But that's not to say that you may not want to figure out how many balloons you need to fill up the big balloon net you rented for your 5-year-old's party.

The example in this section involves much larger spheres — a couple of planets in fact — but I do my best to keep your feet on the ground. Figure 18-3 shows you a sphere.

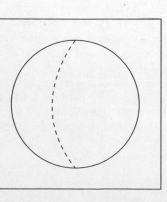

Figure 18-3: Basketballs, globes, planets, and sometimes oranges are spheres.

FUN FACT

Hunting George Dantzig

George Dantzig was born in 1914. As a graduate student, he arrived late to his statistics class one day. He hurriedly copied down the two problems written on the board, assuming they were homework. He did them and turned them in. His excited professor called to tell him that he had solved two famous unsolved problems in statistics!

Dantzig went on to make more of a name for himself when he was hired by the United States Air Force during World War II to find a way to distribute weapons, men, and supplies to various fronts. He developed "linear programming," a process which has had numerous other applications since then.

✔ The diameter of Jupiter is about 88,640 miles, and the diameter of the Earth is about 7,920 miles. How many Earths could fit into Jupiter?

The volume of a sphere (assume that they're both close enough to spheres) is found with $V = \frac{4}{3} \pi \cdot r^3$, so all you need are the measures of the radius of the Earth and the radius of Jupiter. Dividing the diameters by 2, the radius of Earth is about 3,960 miles and the radius of Jupiter is about 44,320 miles.

Number of Earths in Jupiter = Volume of Jupiter ÷ Volume of Earth

$$\text{Number} = \frac{4}{3} \pi \cdot 44,320^3 \div \frac{4}{3} \pi \cdot 3,960^3$$

$$= 44,320^3 - 3,960^3$$

$$\approx 1,400$$

Fourteen hundred Earths could fit into Jupiter. Gives you a better idea of just how big Jupiter is!

Rolling around cylinders

A cylinder has circles for the two bases. The height is the distance measured perpendicularly between the two bases. When cylinders are discussed, you can assume that the bases are above and below one another, as in Figure 18-4 — the cylinder isn't on a slant.

Figure 18-4:
A rolling pin without handles is a cylinder, as are soup cans and soda cans.

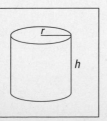

✔ When a soup can manufacturing company increased the height of a can by 3 inches, the volume increased by 12π cubic inches (that's about 37.7 cubic inches). What were the dimensions of the can before and after the increase?

At first, it doesn't seem like there's enough information to answer this question. But the relationship between the dimensions of a cylinder are such that this is a completely solvable problem.

The volume of a cylinder is found with: $V = \pi r^2 h$. The πr^2 part is the area of the circular base, and the h is the height.

The equation reads just like the statement:

The height of the cylinder is increased by 3 inches: $\pi r^2 (h + 3) = V$

The original volume increases by 12π: $\pi r^2 h + 12\pi = V$

$$\pi r^2 (h + 3) = \pi r^2 h + 12\pi$$

Distribute on the left.

$$\pi r^2 h + 3\pi r^2 = \pi r^2 h + 12\pi$$

Subtract the $\pi r^2 h$ from each side.

$$3\pi r^2 = 12\pi$$

Divide each side by 3π.

$$r^2 = 4$$

The radius is 2 inches.

$$r = 2$$

The height can actually be any value. You could substitute any number for the height, and as long as $r = 2$, you'll always get a *true statement,* such as $\pi = \pi$ or $0 = 0$. This means that if the radius is 2 inches, a 3-inch increase in height always has the same change in volume.

Making Up Mixtures

Mixture problems can take on many different forms. There are the traditional types where you can actually mix one solution and another, such as water and antifreeze. There are the types where different solid ingredients are mixed, such as in a salad bowl or candy dish. Another type is where different investments at different interest rates are mixed together. I lumped all these types of problems together because they use basically the same process to solve them.

Drawing a picture helps with all mixture problems. The same picture can work for all: liquid, solid, and investments. Figure 18-5 shows three sample containers — two added together to get a third (the mixture). In each case, the containers are labeled with the "quality" and "quantity" of the contents. These two things get multiplied together before adding. The *quality* is the strength of the antifreeze or the percentage of the interest or the price of the ingredient. You can use the same picture for the containers in every mixture problem or you can change to bowls or boxes. It doesn't matter — you just want to visualize the way the mixture is going together.

Figure 18-5: Visualizing containers can help with mixture problems.

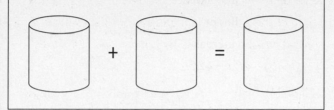

Mixing up solutions

The most popular solutions-type problem is where you mix water and antifreeze. So, to satisfy the car owners, a water/antifreeze problem follows here. You can also solve a problem involving my favorite drink: chocolate milk. When the liquids are mixed, the strengths of the two liquids average out.

✔ How many quarts of 80% antifreeze have to be added to 8 quarts of 20% antifreeze to get a mixture of 60% antifreeze?

First label your containers. The first would be labeled 80% on the top and x on the bottom. (I don't know how many quarts have to be added.) The second container would be labeled with 20% on the top and 8 quarts on the bottom. The third container, which represents the final mixture, would be labeled 60% on top and $x + 8$ quarts on the bottom. To solve this, multiply each "quality" or percentage strength of antifreeze times its "quantity" and put these in the equation.

$$80\% \times x \text{ quarts} + 20\% \times 8 \text{ quarts} = 60\% \times (x + 8) \text{ quarts}$$

$$0.8x + 0.2(8) = 0.6(x + 8)$$

$$0.8x + 1.6 = 0.6x + 4.8$$

Subtracting $0.6x$ from each side and subtracting 1.6 from each side, I get $0.2x = 3.2$.

$x = 16$ quarts of 80% antifreeze have to be added.

> ✔ How much milk and chocolate syrup have to be mixed together to get a 12 ounce glass of chocolate milk that is 30% chocolate syrup?

The percentages are in terms of the percent of chocolate syrup, so the chocolate syrup alone is 100% and the milk alone is 0%. Let x represent the number of ounces of milk.

Your first container should be labeled 0% chocolate syrup on top and x ounces on the bottom. The second container should be labeled 100% on top and $12 - x$ ounces on the bottom. You don't want the glass to over-flow, so and the chocolate syrup takes up space, so you have to subtract the syrup from the total 12 ounces. The third container, the mixture, should have 30% on top and 12 ounces on the bottom.

$$0\% \times x\, oz + 100\% \times (12 - x)\, oz = 30\% \times 12\, oz$$

$$0 + 1(12 - x) = 0.3(12)$$

$$12 - x = 3.6$$

$x = 8.4$ ounces of milk; to that I add 3.6 ounces of chocolate syrup.

You can use the liquid mixture rules with salad dressings, mixed drinks — all sorts of sloshy concoctions.

Tossing in some solid mixtures

You also have many opportunities for mixing solids: mixing the dry ingredients for a cake, tossing a green salad, and making that good old raisins and peanuts mixture, gorp (I fear I'm in a food frenzy). This section demonstrates how to mix solid objects using algebra.

Do you ever buy a can of mixed nuts? I always pick out the cashews. Do you wonder why there seems to be so few of your favorite type and so many peanuts? Well, some types of nuts are more expensive than others, and some are more popular than others. The nut folks (notice I resisted temptation there) take these factors into account when they devise the proportions for a mixture that is both desirable and affordable.

> ✔ How many pounds of cashews that cost \$5.50 per pound should be mixed with 3 pounds of peanuts that cost \$2.00 per pound to create a mixture that costs \$3.00 per pound? (You can use this formula to save your budget for your next big party.)

Using containers makes sense here. Let x represent the number of pounds of cashews. The quality is the cost of the nuts and the quantity is the number of pounds. The first container should have \$5.50 on the top and x pounds on the bottom. The second container should be \$2.00

on top and 3 pounds on the bottom. The third container, with the mixture, should have $3.00 on top and $x + 3$ on the bottom.

$$5.50x + 2.00 (3) = 3.00 (x + 3)$$

$$5.50x + 6.00 = 3.00x + 9.00$$

Subtracting $3.00x$ from each side and 6.00 from each side, you get $2.50x = 3.00$.

$x = \dfrac{3.00}{2.50} = 1.2$ pounds of cashews mixed with 3 pounds of peanuts to create a mixture of 4.2 pounds of nuts that costs $3.00 per pound.

Investigating investments and interest

You can invest your money in a safe CD or savings account and get one interest rate. You can also invest in riskier ventures and get a higher interest rate, but you risk losing money. Most financial advisors suggest that you diversify — put some money in each type of investment — to take advantage of each investment's good points.

Use the simple interest formula in each of these to simplify (in practice, financial institutions are more likely to use the compound interest formula). With simple interest, the interest is figured on the beginning amount only. Compound interest is figured on the changing amounts as the interest is periodically added into the original investment.

✔ Khalil had $20,000 to invest last year. He invested some of this money at $3\frac{1}{2}$% interest and the rest at 8% interest. His total earnings in interest, for both of the investments, were $970. How much did he have invested at each rate?

Use containers again. Let x represent the amount of money invested at $3\frac{1}{2}$%. The first container has $3\frac{1}{2}$% on top and x on the bottom. The second container has 8% on top and $20,000 - x$ on the bottom. The third container, the mixture, has $970 right in the middle. That's the result of multiplying the mixture percentage times the total investment of $20,000. You don't need to know the mixture percentage — just the result.

$$3\tfrac{1}{2}\%(x) + 8\%(20,000 - x) = 970$$

$$0.035 (x) + 0.08 (20,000 - x) = 970$$

$$0.035x + 1600 - 0.08x = 970$$

Subtract 1600 from each side and simplify on the left side.

$$-0.045x = -630$$

$x = 14,000$; that means that $14,000 was invested at $3\frac{1}{2}$% and the other $6,000 was invested at 8%.

✔ Kathy wants to withdraw only the interest on her investment each year. She's going to put money into the account and leave it there, just taking the interest earnings. She wants to take out and spend $10,000 each year. If she puts $\frac{2}{3}$ of her money where it can earn 5% interest and the rest at 7% interest, how much should she put at each rate to have the $10,000 spending money?

Let x represent the total amount of money Kathy needs to invest. The first container has 5% on top and $\frac{2}{3}x$ on the bottom. The second container has 7% on top and $\frac{1}{3}x$ on the bottom. The third container, or mixture, has $10,000 in the middle; this is the result of the "mixed" percentage and the total amount invested.

$$5\% \times \left(\frac{2}{3}x\right) + 7\% \times \left(\frac{1}{3}x\right) = 10,000$$

$$0.05\left(\frac{2}{3}x\right) + 0.07\left(\frac{1}{3}x\right) = 10,000$$

Change the decimals to fractions and multiply.

$$\frac{1}{30}x + \frac{7}{300}x = 10,000$$

Find a common denominator and add the coefficients of x.

$$\frac{17}{300}x = 10,000$$

Divide each side by $\frac{17}{300}$.

$$x \approx 176,470.59$$

Kathy needs over $176,000 to invest. Two-thirds $\left(\frac{2}{3}\right)$ of it, about $117,647, has to be invested at 5% and the rest, about $58,824, at 7%.

Going for the green: Money

Money is everyone's favorite topic. It's something everyone can relate to. It's a blessing and a curse. When you're combining money and algebra, you have to consider the number of coins or bills and their worth or denomination. Other situations involving money can include admission prices, prices of different pizzas in an order, or any commodity with varying prices.

Handling coins and bills

For the purposes of this book, U.S. coins and bills are used in the examples in this section. I don't want to get fancy by including other countries' currency.

✔ A bank teller has five times as many quarters as dimes, three more nickels than dimes, and two fewer than nine times as many pennies as dimes. If she has $15.03 in coins, how many of them are quarters?

The *containers* work here, too. There will be four of them added together: dimes, quarters, nickels, and pennies. The quality is the value of each coin. Every coin count refers to dimes in this problem, so let the number of dimes be represented by x and compare everything else to it.

The first container would contain dimes. Put 0.10 on top and x on the bottom. The second container contains quarters; put 0.25 on top and $5x$ on the bottom. The third container contains nickels; so put 0.05 on top and $x + 3$ on the bottom. The fourth container contains pennies; put 0.01 on top and $9x - 2$ on the bottom. The *mixture* container, on the right, has $15.03 right in the middle.

$$0.10\,(x) + 0.25\,(5x) + 0.05\,(x + 3) + 0.01\,(9x - 2) = 15.03$$

$$0.10x + 1.25x + 0.05x + 0.15 + 0.09x - 0.02 = 15.03$$

$1.49x + 0.13 = 15.03;$ simplifying on the left.

$1.49x = 14.90;$ subtracting the 0.13.

$x = 10$

Because x is the number of dimes, there are 10 dimes, five times as many or 50 quarters, three more or 13 nickels and two less than nine times or 88 pennies. The question was, "How many quarters?" There were 50 quarters; the other answers are used to check to see if this comes out correctly.

Confronting other commodities

Other mixture-type problems that deal with money can take on many forms. The typical problems deal with ticket sales or box office receipts. I do one of those here, plus a retail sales problem.

✔ The total day's receipts at an amusement park were $27,500. Adult tickets were $15 each, and children's tickets were $5 each. How many children were at the park that day, if the total number of paying guests was 3,500?

Let x represent the number of children that came to the park that day. Add the price of the children's tickets times the number of children ($5x$) to the price of the adult tickets (15) times the number of adults ($3,500 - x$) to get the $27,500 total.

$$5x + 15\,(3500 - x) = 27,500$$

$$5x + 52,500 - 15x = 27,500$$

$$52,500 - 10x = 27,500$$

$$-10x = -25,000$$

$$x = 2,500$$

There were 2,500 children at $5 each and 1,000 adults at $15 each.

Playing the averages

When you are asked for the average grade or average height, you usually assume that it's the average where you add up all of the scores or heights and then divide by the number of things that were added together. This is called the *mean* average.

There are actually two other frequently used averages that give "good" answers to the question, "What is usual? What's in the expected?" The other two averages are the median and the mode. The *median* is the actual middle score, if you put scores in order from smallest to largest or largest to smallest. The *mode* is the most frequently occurring score.

Consider the numbers: 1, 1, 2, 2, 4, 4, 4, 4, 5. The mean average is: 27 ÷ 9 = 3; the median is the middle score, or 4. The mode is the most frequently used score, also 4. Each of these averages is "right." The different averages are used in different situations. Sometimes one of them, and not the others, gives a better answer to "What's expected?"

✔ A retailer had a sale on shoes last week. Selected shoes sold for $40, $60, or $70 a pair. The owner sold twice as many $60 shoes as $70 shoes and five times as many $40 shoes as $60 shoes. His revenue for these sale items was a total of $7080. How many of each price did he sell?

Let x represent the number of pairs of $70 shoes. There is no one item that all the others are being compared to, but choosing these shoes allows you to avoid fractions. It *can* be done letting the variable represent another price. I've just chosen what I consider to be the easiest.

If x represents the number of pairs of $70 shoes, then twice that is $2x$, the number of pairs of $60 shoes. Five times the number of $60 shoes would be $5(2x) = 10x$ pairs of $40 shoes.

The first *container* has x and $70, the second has $2x$ and $60, and the third has $10x$ and $40. The sum of the products is $7,080.

$$70(x) + 60(2x) + 40(10x) = 7,080$$

$$70x + 120x + 400x = 7,080$$

$$590x = 7,080$$

$x = 12$; he sold 12 pairs of $70 shoes, $2(12) = 24$ pairs of $60 shoes, and $10(12) = 120$ pairs of $40 shoes.

Going the Distance

You travel, I travel, everybody travels, and at some point everybody asks "Are we there yet?" Algebra can't answer that question for you, but it can help you estimate how long it takes to get there — wherever "there" is.

The distance formula, $d = rt$, says that **distance** is equal to the *rate* of speed multiplied by the *time* it takes to get from the starting point to the destination. You can apply this formula and its variations to determine how long, how far, and how fast you travel. In this section, these applications are divided into the distance plus distance problems and distance equals distance problems. They have common patterns and methods for solving them. The *distance plus distance* problems occur when two objects are going in opposite directions or toward each other and you use the sum of their distances. *Distance equals distance* problems are when the objects go in the same direction with the same starting and ending places — they just go at different speeds or for different amounts of time.

Figuring distance plus distance

One of the two basic distance problems involves when one object travels a certain distance, a second object travels another distance, and the two distances are added together. There could be two kids on walkie-talkies, going in opposite directions to see how far apart they'd have to be before they couldn't communicate any more. Another instance would be when two cars leave different cities heading toward each other on the same road and you figure out where they meet.

✔ Deirdre and Donovan are in love and will be meeting in Kansas City to get married. Deirdre boarded a train at noon traveling due east toward Kansas City. Two hours later, Donovan boarded a train traveling due west, also heading for Kansas City, and going at a rate of speed 20 miles per hour faster Deirdre. At noon, they were 1,100 miles apart. At 9:00 p.m., they both arrived at Kansas City. How fast were they traveling?

Distance of Deirdre from Kansas City + Distance of Donovan from Kansas City = 1,100

Rate × Time + Rate × Time = 1,100

Let the speed (rate) of Dierdre's train be represented by r. Donovan's train was traveling 20 miles per hour (mph) faster Dierdre's, so the speed of Donovan's train is $r + 20$.

Let the time traveled by Dierdre's train be represented by t. Donovan's train left 2 hours after Dierdre's, so the time traveled by Donovan's train is $t - 2$.

$$r \cdot t + (r + 20)(t - 2) = 1,100$$

The time is 9 hours for Dierdre's train and $t - 2$ or 7 hours for the Donovan's. Replacing these values in the equation,

$$r \cdot 9 + (r + 20) \cdot 7 = 1,100$$

Now distribute the 7.

$$9r + 7r + 140 = 1,100$$

Combine the two terms with r.

$16r = 960$

Divide each side by 16.

$r = 60$. Dierdre's train is going 60 mph; Donovan's is going $r + 20$ or 80 mph.

Figuring distance equals distance

The other traditional distance problems involve situations where you set one distance *equal* to another distance. When one person has to catch up to another or when a police car has to catch a speeder — these are typical problems where the distances are the same but the speeds and times are different.

✔ Taurique left for work at 7:00 a.m. riding his bicycle at an average rate of 12 miles per hour (mph). At 7:15 a.m., Tanisha noticed that he had forgotten his briefcase and followed him in her car at an average rate of 24 miles per hour. When did Tanisha catch up with Taurique?

In this case, the two distances traveled are the same:

Distance Taurique rides = Distance Tanisha drives, and Taurique's rate × Taurique's time = Tanisha's rate × Tanisha's time

Let t represent the amount of time that Taurique traveled. Because Tanisha didn't leave until 15 minutes later, represent her time by $t - \frac{1}{4}$. Change the 15 minutes into $\frac{1}{4}$ hour because the rates are in miles per *hour*.

$$12\,\text{mph} \times t = 24\,\text{mph} \times \left(t - \frac{1}{4}\right)$$

$$12 \cdot t = 24\left(t - \frac{1}{4}\right)$$

$$12 \cdot t = 24t - 6$$

$$-12t = -6$$

$t = \frac{1}{2}$. It took $\frac{1}{2}$ hour from the time Taurique left until Tanisha caught up with him; she met him at 7:30 a.m.

Righting Right Triangles

Ever since Pythagoras determined how wonderful right triangles and the relationships between their sides are, the applications have been endless. The Pythagorean theorem says that the relationship between the two shorter sides, a and b, and the hypotenuse c — the long side opposite the right angle — is $a^2 + b^2 = c^2$.

Scaling the heights

You can determine the height of many objects using the formulas for right triangles. For example, you can figure out the height of a building or tree using its shadow. You can figure the height of a flagpole or how high up that fellow with the water balloon is. Of course, to discover the height of objects outdoors, you have to assume that trees and buildings are at right angles to the ground. Although this may not always be *exactly* true, it's usually close enough for practical purposes.

✔ Tom is flying his new kite. When Tom has 200 feet of string played out, Don is directly underneath the kite and is 50 feet away from Tom. How high is the kite? Figure 18-6 gives you a visual to work with.

In this problem, the amount of string played out is at an angle; the string forms the hypotenuse of the right triangle. The distance from Tom to Don is the bottom leg of the right triangle. In this problem, I'll let a be that bottom leg and c be the string.

$$a^2 + b^2 = c^2$$

$$50^2 + b^2 = 200^2$$

$$2,500 + b^2 = 40,000$$

$$b^2 = 37,500$$

$$b = \sqrt{37,500} \approx 193.65 \text{ feet in the air}$$

✔ Tom's kite is stuck on the top of a tree. The tree is casting a 30-foot shadow onto the ground. Tom has a yardstick (isn't that handy) which he stands up on the ground, and it casts a shadow 2 feet long. Tom wants to know whether he needs a ladder or a helicopter to get his kite down. Figuring out the height of the tree helps him make this decision.

Figure 18-6:
Kites and
heights.

In this case, two right triangles are formed. One is the tree, its shadow, and the line from the top of the tree to the end of the shadow. The other right triangle is the yardstick, its shadow, and the line from the top of the yardstick to the end of its shadow. The right triangles are similar to each other because the sun is hitting the tops at the same angle. Write a proportion to solve this, using the two shadows, or sides *a*, and the two heights, or sides *b*.

(shadow of tree) ÷ (height of tree) = (shadow of stick) ÷ (height of stick)

Let *x* represent the height of the tree, since that's the only value you don't know yet.

$$\frac{30 \text{ feet}}{x} = \frac{2 \text{ feet}}{3 \text{ feet}}$$

$$\frac{30}{x} = \frac{2}{3}$$

$30 \cdot 3 = 2 \cdot x$ by cross-multiplying.

$90 = 2x$

$x = 45$, so the tree is 45 feet high. Tom had better forget about the kite.

Solving surface distances

Sometimes a right triangle can be used to find a distance when you can't make all the measurements. Such instances might arise when finding the distance across a ravine, lake, or swamp.

✔ A small town wants to build a bridge across the local lake to make it easier to get to the big city. The city is due east, across the lake from the town. If you drive due south from the small town for $1\frac{1}{3}$ miles, you reach a road that goes straight to the city. The distance from where the roads intersect to the big city is $1\frac{2}{3}$ miles. How long must the bridge be?

This is a hard way to draw a right triangle, but there is a right triangle here. The bridge is one leg. The distance going south to the intersection of the roads is another leg. The road from the south intersection to the big city is the hypotenuse.

$$a^2 + b^2 = c^2$$

Let *a* represent the length of the bridge.

$$a^2 + \left(1\frac{1}{3}\right)^2 = \left(1\frac{2}{3}\right)^2$$

$$a^2 + \frac{16}{9} = \frac{25}{9}$$

$$a^2 = \frac{9}{9} = 1$$

So *a* is 1 mile. If they build a bridge that is one mile long, they will save driving the 3 miles to the city. Sounds a bit expensive.

🖝 Two planes leave the airport at the same time. One flies north at an unknown speed, and the other flies east at 400 miles per hour. After 3 hours, the planes are 1,500 miles apart. How fast is the northbound plane traveling?

A right triangle is formed with the legs going to the north and east. The hypotenuse is the distance that they are apart.

$$a^2 + b^2 = c^2$$

Let a represent the distance that the northbound plane has traveled after 3 hours. The b can represent how far the eastbound plane has traveled; this would be 1,200 miles, because $d = rt$, and the plane flew at 400 mph for 3 hours. The c represents the distance apart, or 1500 miles.

$$a^2 + 1,200^2 = 1,500^2$$

$$a^2 = 1,500^2 - 1,200^2$$

$$a^2 = 810,000$$

$$a = 900$$

Because a represents the *distance* traveled in 3 hours, the speed can be determined with $d = rt$.

$$900 = r \cdot 3$$

$$r = 300 \text{ miles per hour.}$$

Going 'Round in Circles

The circle, as Figure 18-7 shows, is a very nice, efficient figure, although using a circular shape isn't always practical. Circles don't fit together well. There are always gaps between them, so they don't make good shapes for fields, yards, or areas shared with others. Even with this problem, though, circles are useful and provide problems involving their area: circular rugs and race tracks; fields and swimming pools.

Figure 18-7:
The diameter is the longest distance across a circle.

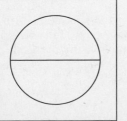

The area of a circle can be determined if you know the radius or the diameter.

- Grace decided to get an 18-foot-diameter, aboveground pool instead of a 12-foot-diameter pool. How much more area will this bigger pool cover?

  The area of a circle is found with: $A = \pi \cdot r^2$.

  The diameter of a circle is twice the radius, so an 18-foot-diameter pool has a radius of 9 feet, and the 12-foot-diameter pool has a 6-foot radius.

  Difference in area = Area of bigger pool – Area of smaller pool

  Difference = $\pi(9)^2 - \pi(6)^2 = 81\pi - 36\pi = 45\pi \approx 141.4$ square feet

- If you have a certain amount of fencing, you can enclose more area in a circular shape than in any other shape. To prove this point, this example asks how much bigger a circular yard enclosed by 314 feet of fencing is than a square yard enclosed in the same amount of fencing.

  The area of a circle is found with $A = \pi r^2$, and the area of a square is found with $A = s^2$. It looks like this will be fairly simple; you just have to find the difference between the two values.

  Difference = Area of Circle – Area of Square

  The challenge comes in when you need the value of r, the radius of the circle and the value of s, the length of a side of the square. You don't have the value of r or the value of s. You just have the distance around the outside called the *perimeter*, and there's a formula for each figure.

  The perimeter of a circle is its *circumference* which is found with $C = 2\pi r$, and the perimeter of a square is found with $P = 4s$.

  If 314 is the circumference of the circle, then $314 = 2\pi r$, or $r = {}^{314}/_{2\pi} \approx 50$.

  So the area of the circle is $A = \pi 50^2 = 2500\pi \approx 7,854$ square feet.

  The perimeter of a square is just 4 times the measure of the side.

  Because 314 is the perimeter of the square, then $314 = 4 \cdot s$, or $s = 78.5$.

  That means that the area of the square is $A = 78.5^2 = 6,162.25$.

  Difference = $7,854 - 6,162.25 = 1,691.75$ square feet. That's quite a bit more area in the circle than in the square.

A goal of this chapter is to provide some real-life situations or problems that you can solve using algebra. Perhaps you saw yourself in some of the examples. Perhaps you could never imagine yourself in any of the situations. In any case, I hope you can use the patterns or models to set up and solve problems similar to one of these, as the need arises.

Chapter 19

Going Visual: Graphing

A picture is worth a thousand words. This is especially true in algebra. Pictures or graphs give you an instant impression of what's happening in a situation or what an equation is representing in space. A *graph* is a drawing that illustrates an algebraic operation or equation in a two-dimensional plane (like a piece of graph paper). A graph allows you to see the characteristics of an algebraic statement immediately. The words needed to describe what you see in a graph may be lengthy and complicated.

Most people are familiar with bar graphs and their rectangles standing on end, which often depict test scores. Pie graphs — a circle with wedges of varying sizes — are good for showing proportions of a whole, such as how money is spent, and line graphs are great for showing the ups and downs of the stock market and for showing activity over time.

The graphs in algebra are unique because they reveal relationships that you can use to model a situation: A line can model depreciation of value; parabolas can model daily temperature; or a flat, *S*-shaped curve can model the number of people infected with the flu. All these and other models are useful for showing what's happening and predicting what can happen in the future.

Algebraic equations match up with the graph. With algebraic operations and techniques applied to equations to make them more usable, the equations can be used to predict, project, and figure out various problems.

Graphing Is Good

Consider the three ways of expressing the same thing in each of the following examples: in words, in an algebraic equation, and in a graph.

> ✔ All the pairs of numbers that add up to 10.
>
> ✔ $x + y = 10$
>
> ✔ The graph shown in Figure 19-1.

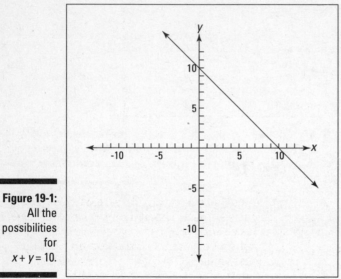

Figure 19-1:
All the possibilities for $x + y = 10$.

And, again:

> ✔ All the pairs of numbers you get when you choose a number and then get the second number by subtracting the first number from its square.
>
> ✔ $y = x^2 - x$
>
> ✔ The following graph (see Figure 19-2).

The algebraic equation describes the situation in a more concise manner than the wordy description. The graph, however, gives you a better idea of what is being described than the words or the equation.

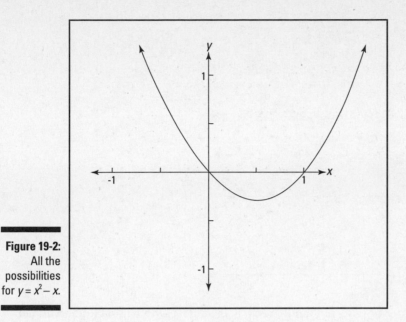

Figure 19-2:
All the
possibilities
for $y = x^2 - x$.

Grappling with Graphs

The cartoons in the newspaper often show a worried businessperson pointing to a graph full of ups and downs — usually the punch line involves a huge drop in sales. As entertaining as these cartoons may be, they also cut right to a major usefulness of graphs. Graphs give instant information on what the lowest value is and what the highest value is. They give information on trends, patterns, and the current status. Another great application is to put two graphs in the same picture to compare them.

In the simplest terms, a graph in algebra is drawn on a *coordinate plane* — two lines that cross one another at right angles to form four sections or *quadrants*. The two lines, or *axes* (pronounced ax-eez), are number lines usually marked with the integers (positive and negative whole numbers and zero). The positives go upward on the vertical axis and to the right on the horizontal axis. Figure 19-3 shows a coordinate plane. The line going left and right — the horizontal line — is the *x-axis* and the vertical line going up and down is the *y-axis*.

The little marks on the axes are called *tick marks.* They are all uniformly spaced (like the tick-tocks of a clock are the same time apart) and are usually labeled with the integers, negative to positive, left to right, and downward to upward, with zero in the middle — at the point where the axes meet.

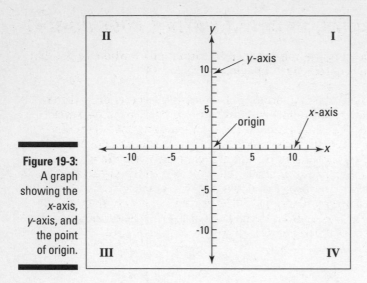

The four quadrants are numbered I, II, III, and IV, with capital Roman numerals starting with the upper right quadrant and going counter-clockwise. The reason for this is simple: It's tradition. Table 19-1 illustrates the quadrants and their names and positions.

Making a point

You do the type of point finding needed to do graphing when you find the whereabouts of Peoria, Illinois, at G7 on a road atlas. You move your finger so it's down from the G and across from the 7. Graphing in algebra is just a bit different because numbers replace the letters, and you start in the middle at the point of origin.

Points are dots on a piece of paper or blackboard that represent positions or places with respect to the axes — vertical and horizontal lines — of a graph. The coordinates of a point tell you where it is in the graph.

The axes of an algebraic graph are usually labeled with integers, but they can be labeled with any rational numbers, as long as the numbers are the same distance apart from each other, such as the one-quarter distance between $\frac{1}{4}$, $\frac{1}{2}$, and $\frac{3}{4}$.

The points you put *in* the graph don't have to be integers or line up with the axes. They can be in the spaces between the lines. That way, they can represent any number. The points in the spaces are just estimates of numbers, though. You could tell from the graph that a point is between 2 and 3, but you'd have a harder time deciding whether that point is 2.5 or 2.6.

Ordering pairs, or coordinating coordinates

To actually put a point in a graph, you need information on where to put that point. That's where ordered pairs come in.

An *ordered pair* is a set of two numbers called *coordinates* that are written inside a parenthesis with a comma separating them. Some examples are: (2, 3), (−1, 4), and (5, 0). This particular notation indicates that the order matters: The first number, or *x-coordinate,* tells you the point's position with respect to the *x*-axis — how far to the left or right from the point of origin — and the second number, or *y-coordinate,* tells you the point's position with respect to the *y*-axis — how far up or down from the point of origin. For example, the point for the ordered pair (3, 2) is three moves to the right of the origin, and two moves up from there. Look at Figure 19-4 to see where the points are for several ordered pairs.

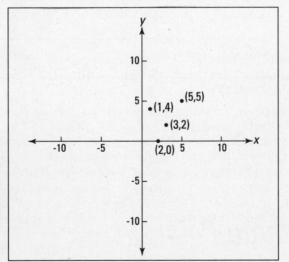

Figure 19-4:
Coordinates and their points on a graph.

Everything starts at the *origin* — the intersection of the two axes. The ordered pair for the origin is (0,0). The numbers in this ordered pair tell you that the point didn't go left, right, up, or down. Its position is at the starting place.

Notice that point (2,0) lies right on the *x*-axis. Whenever 0 is a coordinate within the ordered pair, then the point *must be* located on an axis.

Table 19-1 shows the names of the quadrants, their positions in the coordinate plane, and characteristics of coordinate points in the various quadrants.

Table 19-1		Quadrants	
Quadrant	*Position*	*Coordinate Values*	*How to Plot*
Quadrant I	Upper right side	positive, positive	Move right and up.
Quadrant II	Upper left side	negative, positive	Move left and up.
Quadrant III	Lower left side	negative, negative	Move left and down.
Quadrant IV	Lower right side	positive, negative	Move right and down.
	Right axis	positive, 0	Move right and sit on the x-axis.
	Left axis	negative, 0	Move left and sit on the x-axis.
	Upper axis	0, positive	Move up and sit on the y-axis.
	Lower axis	0, negative	Move down and sit on the y-axis.

The actual graphing on the coordinate plane involves placing dots for points in their correct position. Sometimes the points are sitting there, all alone, like an unused page in a child's "connect the dots" book. When the dots get connected, they sometimes form lines, U-shaped curves, or even more exciting pictures. It depends on the equation that the points come from.

Actually Graphing Points

To plot a point, look at the coordinates — the numbers in the parenthesis. The first number tells you which way to move, horizontally, from the origin. Place your pencil on the origin and move right if the first number is positive; move left if the first number is negative. Next, from that position, move your pencil up or down: up if the second number is positive and down if it's negative.

The following points are graphed in Figure 19-5. The letters serve as names of the points so you can compare their coordinates.

A: (9,0)	B: (7,4)	C: (3,8)	D: (0,7)	E: (-2,2)
F: (-8,0)	G: (-5,-3)	H: (0,-3)	J: (3,-2)	K: (8,-7)

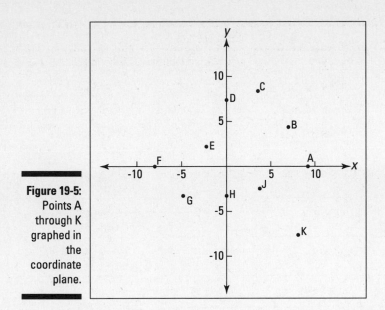

Figure 19-5:
Points A
through K
graphed in
the
coordinate
plane.

Graphing a Line

Lines are one of the most basic and most useful things to graph in algebra. You can use them to represent how your income fluctuates or how a distance from a point changes. They can represent how a piece of machinery depreciates. So, lines are useful, and they're easy to deal with, too. What more could you ask?

Dots or points scattered all over the place with no apparent shape don't usually mean anything. In algebra, it's more common to see points arranged with an equation that gives them something in common. The simplest pattern is a straight line. Line up those points!

A *line* (straight line) is the set of all the points on a graph that satisfy a linear equation.

These points are graphed in Figure 19-6: (-2,12), (-1,11), (0,10), (1,9), (2,8), (3,7), (4,6), (5,5), (6,4), (7,3), (8,2), (9,1), (10,0), (11,-1), (12,-2), (13,-3).

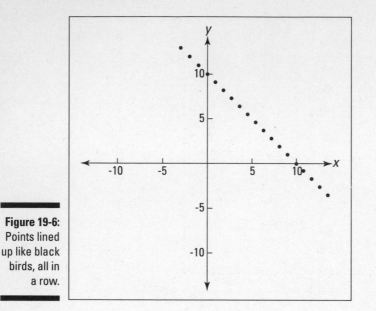

Figure 19-6:
Points lined
up like black
birds, all in
a row.

Rather than list all the millions of points that seem to lie along the same line, you can write an equation that expresses the relation between the points. In this case, the relation is $x + y = 10$. The coordinates in each pair add up to 10. But what the graph doesn't show is that all the points are *not* integers — points such as $\left(1\frac{1}{2}, 8\frac{1}{2}\right)$ that fit the pattern and lie on the line. By connecting all the points to form a line, you include all the fractions between the points.

This equation, $x + y = 10$, says that anything adding up to 10 gives you a point on the graph. This includes fractions, decimals, positives, and negatives. What was at first many points or values that worked in the equation are now an *infinite* number of points.

Figure 19-7 shows you how points look when they're connected to form a line.

To graph a line, you only need two points. A rule in geometry says that only one line can go through two particular points. Even though only two points are needed to graph a line, it's a good idea to graph at least three points to be sure that you graphed the line correctly.

In graphing, three is much better than two. If you get one of two points in the wrong place in a graph, you probably won't notice that the line is wrong. However, if you get one of three points in the wrong place, you're more likely to notice that your line isn't straight. Plotting three points is a good check.

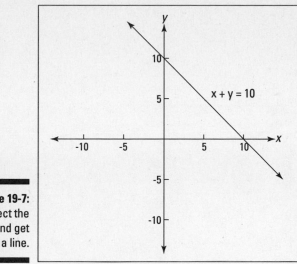

Figure 19-7:
Connect the
dots and get
a line.

Graphing the equation of a line

An equation whose graph is a straight line is said to be *linear*. The equation has a standard form of $ax + by = c$, where x and y are variables and a, b, and c are real numbers. The equation of a line usually has an x or a y, often both, which refer to all the points (x, y) that make the equation true. The x and y each have a power of one (if the powers were higher or lower, the graph would curve).

When graphing a line, you can find some pairs of numbers that make the equation true and then connect them. *Connect the dots!*

What does the equation of a line look like? It looks like any of the following examples. Notice that the first three equations are written in the standard form, and the fourth has you solve for y. The last two have only one variable; this happens with horizontal and vertical lines.

$$x + y = 10 \qquad 2x + 3y = 4$$
$$-5x + y = 7 \qquad y = \frac{1}{2}x + 3$$
$$x = 3 \qquad y = -2$$

Whenever you have an equation where y equals a number, then you have a horizontal line going through all those y values. Conversely, if you have an equation where x equals a number, all the x values are the same, and you have a vertical line. Horizontal lines are all parallel to the x-axis. Their equations look like: $y = 3$ or $y = -2$. Vertical lines are all parallel to the y-axis. Their equations all look like: $x = 5$ or $x = -11$. Figure 19-8 is a graph of $y = 4$, using four points: $(-4,4)$, $(0,4)$, $(1,4)$, and $(3,4)$.

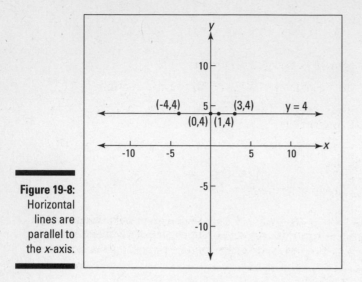

Find a point on the line $x - y = 3$ using the following steps.

1. **Choose a random value for one of the variables, either x or y.**

   Faced with finding points that lie on the line $x - y = 3$, you can let $x = 8$, so $8 - y = 3$.

2. **Solve for the value of the other variable.**

   Subtract 8 from each side to get $-y = -5$

   Multiply each side by -1 to get $y = 5$.

   Remember that you can change the looks of the equation without changing the graph of the line by multiplying or dividing each side by -1.

   (For a review of solving linear equations, go to Chapter 12.)

3. **Write an ordered pair for the coordinates of the point.**

   So your first ordered pair is (8,5).

You find more ordered pairs by choosing another number to substitute for *x* or *y*.

For more of a challenge, find points that lie on the line $2x + 3y = 12$. The multipliers (2 and 3) make this one just a little trickier. You may find one or two points fairly easily, but others could be more difficult because of fractions. A good plan in a case like this is to solve for *x* or *y* and then plug in numbers, as these steps do:

1. Solve the equation for one of the variables.

   Solving for *y* in the sample problem $2x + 3y = 12$ you get:

$$3y = 12 - 2x$$
$$y = \frac{12 - 2x}{3}$$

With multipliers involved, you get a fraction.

2. **Choose a value for the other variable and solve the equation.**

   Try to pick values so that the number in the numerator is divisible by the first variable's multiplier.

   In the sample, 3 is the multiplier for y, so let $x = 3$. Solving the equation,
   $$y = \frac{12 - 2 \cdot 3}{3} = \frac{6}{3} = 2$$
   So, the point (3,2) lies on the line.

Finding the points that lie on the line $x = 4$ may look like a really tough assignment, with only an x showing in the equation. But this actually makes the whole thing much easier. You can write down anything for the y value, as long as x is equal to 4.

Some points are: (4,9), (4,–2), (4,0), (4,3.16), (4,–11), (4,4).

Notice that the 4 is *always* the first number. The point (4,9) is not the same as the point (9,4). Remember that the order counts in ordered pairs.

Graphing these points, gives you a nice, vertical line, as Figure 19-9 shows. If all the y-coordinates are the same point, the line is — you guessed it — horizontal.

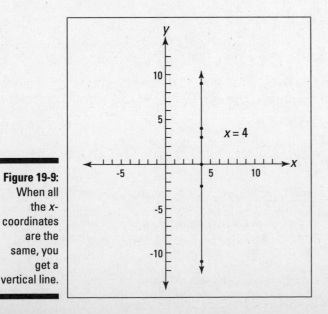

Figure 19-9:
When all the x-coordinates are the same, you get a vertical line.

Investigating Intercepts

The *intercept* of a line is a point where the line crosses an axis. Unless a line is vertical or horizontal, it crosses both the x and y axes, and so has two intercepts — an x-intercept and a y-intercept. Horizontal lines just have a y-intercept, and vertical lines just have an x-intercept. The exceptions are when the horizontal line is actually the x-axis or the vertical line is the y-axis.

Intercepts are quick and easy to find and can be a big help when graphing. The reason for this is that one of the coordinates of every intercept is a zero. Zeros in equations cut down on the numbers and the work, and it's nice to take advantage of this when you can.

The *x-intercept* of a line is where the line crosses the x-axis. To find the x-intercept, let the y in the equation equal zero and solve for x.

✔ To find the x-intercept of the line $4x - 7y = 8$, let $y = 0$ in the equation.

Then $4x - 0 = 8$, $4x = 8$, and $x = 2$.

The x-intercept of the line is $(2,0)$: The line goes through the x-axis at that point.

The *y-intercept* of a line is where the line crosses the y-axis. To find the y-intercept, let the x in the equation equal zero and solve for y.

✔ Find the y-intercept of the line $3x - 7y = 28$

Let $x = 0$ in the equation.

Then $0 - 7y = 28$, $-7y = 28$, and $y = -4$.

The y-intercept of the line is $(0,-4)$.

As long as you're careful when graphing the x- and y-intercepts, and get them on the correct axes, these are sometimes all you need to graph a line.

Sighting the Slope

The *slope* of a line is a number that describes the steepness and direction of the graph of the line. The slope is a positive number if the line moves upward from left to right. The slope is a negative number if the line moves downward from left to right. The steeper the line, the greater the absolute value of the slope (the farther the number is from zero).

Knowing the slope of a line beforehand helps you graph the line. You can find a point on the line and then use the slope to graph it. A line with a slope of 6 goes up steeply. If you know what the line should look like as far as going up or down — information you get from the slope — it's easier to graph it correctly.

The value of the slope is important when the equation of the line is used in modeling situations. For instance, in equations representing the cost of so many items, the value of the slope is called the *marginal cost*. In equations representing depreciation, the slope is the *annual depreciation*.

Figure 19-10 shows some lines with their slopes. The lines are all going through the origin just for convenience.

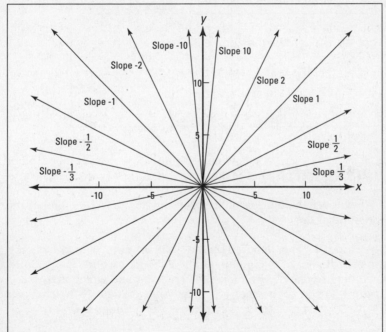

Figure 19-10:
Pick a
line — see
its slope.

What about a horizontal line — one that doesn't go upward or downward? A horizontal line has a zero slope. A vertical line has no slope; the slope is *undefined*.

Figure 19-11 shows graphs of lines that have a zero slope or undefined slope.

One way of referring to the slope, when it's written as a fraction, is *rise over run*. If the slope is $\frac{3}{2}$, it means that for every 2 units the line *runs* along the x-axis, it *rises* 3 units along the y-axis. A slope of $\frac{-1}{8}$ indicates that as the line runs 8 units horizontally, parallel to the x-axis, it drops (*negative rise*) 1 unit vertically.

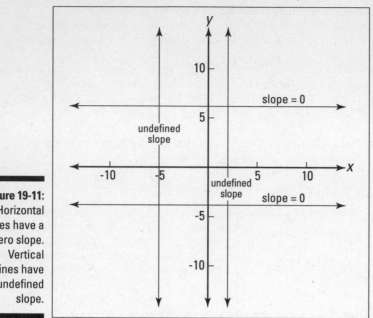

Figure 19-11:
Horizontal
lines have a
zero slope.
Vertical
lines have
undefined
slope.

Formulating slope

If you know two points on a line, you can compute the number representing the slope of the line.

The slope of a line, denoted by the small letter *m*, is found when you know the coordinates of two points on the line, (x_1, y_1) and (x_2, y_2). Subscripts are used here to identify which is the first point and which is the second point. There's no rule as to which is which; you can identify them anyway you want. It's just a good idea to identify them to keep things in order.

$$m = \frac{y_2 - y_1}{x_2 - x_1}$$

This would come out to be the same slope if you subtract in the opposite order

$$m = \frac{y_1 - y_2}{x_1 - x_2}$$

You just can't mix them and do $(x_1 - y_2)$ over $(x_2 - y_1)$.

Now, you can try it with the following examples.

- ✔ Find the slope of the line going through (3,4) and (2,10).

  Let (3,4) be (x_1,y_1) and (2,10) be (x_2,y_2).

  Substitute into the formula:

  $$m = \frac{y_2 - y_1}{x_2 - x_2} = \frac{10 - 4}{2 - 3}$$

  Simplify: $m = \frac{6}{-1} = -6$

  This line is pretty steep as it falls from left to right.

- ✔ Find the slope of the line going through (4,2) and (–6,2).

  Let (4,2) be (x_1,y_1) and (–6,2) be (x_2,y_2).

  Substitute into the formula.

  $$m = \frac{y_2 - y_1}{x_2 - x_2} = \frac{2 - 2}{-6 - 4}$$

  Simplify: $m = \frac{0}{-10} = 0$

  These points are both two units above the *x*-axis and form a horizontal line. That's why the slope is 0.

- ✔ Find the slope of the line going through (2,4) and (2,–6).

  Let (2,4) be (x_1,y_1) and (2,–6) be (x_2,y_2).

  Substitute into the formula.

  $$m = \frac{y_2 - y_1}{x_2 - x_2} = \frac{-6 - 4}{2 - 2}$$

  Simplify: $m = \frac{-10}{0}$

Oops! You can't divide by zero. There is no such number. The slope doesn't exist or is *undefined*. These two points are on a vertical line.

Watch out for these two common errors when working with the slope formula:

- ✔ Be sure that you subtract the *y* values on the *top* of the division formula. A common error is to subtract the *x* values on the top.

- ✔ Be sure to keep the numbers in the same order when you subtract. Decide which point is first and which point is second. Then take the second *y* minus the first *y* and the second *x* minus the first *x*. Don't do the top in a different order from the bottom.

Combining slope and intercept

An equation of a single line can take many forms. Just as you can solve for one variable or another in a formula, you can solve for one of the variables in the equation of a line. This "instant information" can help you find the points to graph the line or find the slope of a line.

A common and popular form of the equation of a line is the *slope-intercept* form. It's called this because the slope of the line and the y-intercept of the line are obvious on sight. When a line is written $6x + 3y = 5$, you can find points by plugging in numbers for x or y and solving for the other coordinate. By using methods for solving linear equations (Chapter 12), the same equation can be written $y = -2x + \frac{5}{3}$, which tells you that the slope is -2 and the place the line crosses the y axis (the y-intercept) i s $\left(0, \frac{5}{3}\right)$.

 Where y and x represent points on the line, m is the slope of the line, and b is the y-intercept of the line, the *slope-intercept* form is

$$y = mx + b$$

In every case here, the equation is written in the slope-intercept form. The coefficient of x is the slope of the line and the constant is the y-intercept.

- ✔ $y = 2x + 3$: The slope is 2; the y-intercept is (0,3).
- ✔ $y = \frac{1}{3}x - 2$: The slope is $\frac{1}{3}$; the y-intercept is (0,−2).
- ✔ $y = 7$: The slope is 0; the y-intercept is (0,7). You can read this equation as being $y = 0 \cdot x + 7$.

Getting to the slope-intercept form

If the equation of the line isn't already in the slope-intercept form, solving for y can change the equation to slope-intercept form. The following steps show you how.

Put the equation $5x - 2y = 10$ in slope-intercept form.

1. **Get the y term by itself on the left.**

   Subtract $5x$ from each side to get the y term alone: $-2y = -5x + 10$

2. **Solve for y.**

   Divide each side by -2 and simplify the two terms on the right.

$$\frac{-2y}{-2} = \frac{(-5x + 10)}{-2}$$

$$y = \frac{-5x}{-2} + \frac{10}{-2}$$

$$y = \frac{5}{2}x - 5$$

The slope is $\frac{5}{2}$ and the y-intercept is at (0,–5).

Graphing with slope-intercept

One advantage to having an equation in the slope-intercept form is that graphing the line can be a fairly quick task, as the following examples show.

✔ Graph $y = \frac{3}{2}x + 1$

The slope of this line is $\frac{3}{2}$ and the y-intercept is the point (0,1). First graph the y-intercept. (See Figure 19-12.)

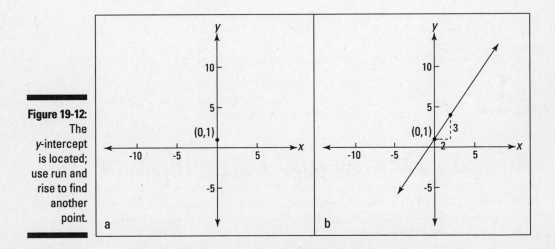

Figure 19-12: The y-intercept is located; use run and rise to find another point.

Then use the *rise over run* interpretation of slope to count spaces to another point on the line. To do this, do the run, or bottom, movement first. In this sketch, move 2 units to the right of (0,1). From there, *rise* or go up 3 units, which should get you to (2,4).

It's sort of like going on a treasure hunt: "Two steps to the east; three steps to the north; now dig in!" Only our "dig in" is to put a point there and connect that point with the starting point — the intercept. Look at the right-hand side (the b side) of Figure 19-12 to see how it's done.

This is a quick and easy way to sketch a line.

✔ Graph $y = -3x + 2$

First, graph the y-intercept (0,2).

Think of the slope -3 as being the fraction $\frac{-3}{1}$. This way you have a *run* of 1. The *rise* isn't a rise in this case. The 3 is negative, so it's a fall. Connect the intercept (0,2) with the point that you find by moving 1 unit to the right and 3 units down, which should be (1,–1).

Figure 19-13 shows the line $y = -3x + 2$ which has a slope of -3.

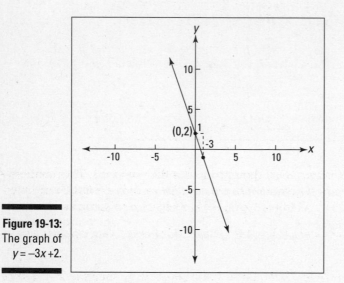

Figure 19-13: The graph of $y = -3x + 2$.

Marking Parallel and Perpendicular Lines

The slope of a line gives you information about a particular characteristic of the line. It tells you if it's steep or flat and whether it's rising or falling as you read from left to right. The slope of a line can also tell you if one line is parallel or perpendicular to another line. Figure 19-14 shows parallel and perpendicular lines.

Parallel lines never touch. They are always the same distance apart and never share a common point. They have the same slope.

Perpendicular lines form a 90-degree angle (a right angle) where they cross. They have slopes that are *negative reciprocals* of one another. The x and y axes are perpendicular lines.

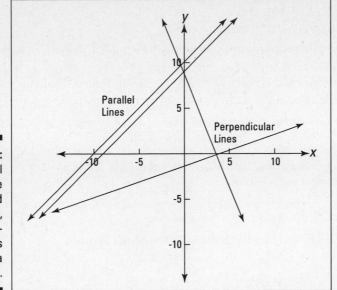

Figure 19-14:
Parallel lines are like railroad tracks, perpendicular lines meet at a right angle.

Two numbers are *reciprocals* if their product is the number 1. The numbers $\frac{3}{4}$ and $\frac{4}{3}$ are reciprocals. Two numbers are *negative reciprocals* if their product is the number –1. The numbers $\frac{3}{4}$ and $-\frac{4}{3}$ are negative reciprocals.

If line ℓ_1 has a slope of m_1, and if line ℓ_2 has a slope of m_2, then the lines are *parallel* if $m_1 = m_2$.

If line ℓ_1 has a slope of m_1, and if line ℓ_2 has a slope of m_2, then the lines are *perpendicular* if $m_1 = -\frac{1}{m_2}$.

The following examples show you how to determine whether lines are parallel or perpendicular by just looking at their slopes.

- ✔ The line $y = 3x + 2$ is parallel to the line $y = 3x - 7$ because their slopes are both 3.
- ✔ The line $y = \frac{-1}{4}x + 3$ is parallel to the line $y = -\frac{1}{4}x + 1$ because their slopes are both $\frac{-1}{4}$.
- ✔ The line $3x + 2y = 8$ is parallel to the line $6x + 4y = 7$ because their slopes are both $\frac{-3}{2}$. Write each line in the slope-intercept form to see this.

$$3x + 2y = 8 \text{ can be written } 2y = -3x + 8 \text{ or } y = \frac{-3}{2}x + 4.$$

$$6x + 4y = 7 \text{ can be written } 4y = -6x + 7 \text{ or } y = \frac{-3}{2}x + \frac{7}{4}.$$

- ✔ The line $y = \frac{3}{4}x + 5$ is perpendicular to the line $y = \frac{-4}{3}x + 6$ because their slopes are negative reciprocals of one another.
- ✔ The line $y = -3x + 4$ is perpendicular to the line $y = \frac{1}{3}x - 8$ because their slopes are negative reciprocals of one another.

Intersecting Lines

If two lines *intersect,* or cross one another, then they do that exactly once and only once. The place they cross is the *point of intersection* and that common point is the only one both lines share. Careful graphing can sometimes help you to find the point of intersection.

The point (5,1) is the point of intersection of the two lines $x + y = 6$ and $2x - y = 9$ because the coordinates make each equation true:

1. **If $x + y = 6$, then substituting the values $x = 5$ and $y = 1$ give you 5 + 1 = 6, which is true.**

2. **If $2x - y = 9$, then substituting the values $x = 5$ and $y = 1$ give $2 \cdot 5 - 1 = 10 - 1 = 9$, which is also true.**

3. **This is the only point that works for both the lines.**

Graphing for intersections

Careful graphing can give you the intersection of two lines. The only problem is that if your graph is even a little off, you can get the wrong answer. Also, if the answer has a fraction in it, it's difficult to figure out what that fraction is. The following examples can help you make some sense of this.

✔ Find the intersection of the lines $3x - y = 5$ and $x + y = -1$.

Look at the graphs in Figure 19-15.

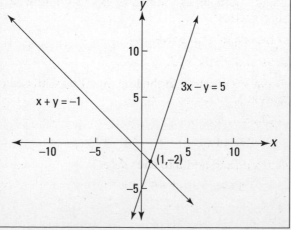

Figure 19-15:
The intersection of two lines at the point (1,−2).

The lines appear to cross at the point (1,–2). Replace the coordinates in the equations to check this out.

If $3x - y = 5$, then substituting the values gives $3 \cdot 1 - (-2) = 3 + 2 = 5$, which is true.

If $x + y = -1$, then substituting the values gives $1 + (-2) = -1$, which is also true.

Graphing is an inexact way to find the intersection of lines. You have to be super careful when plotting the points and lines.

Substituting to find intersections

Another way to find the point where two lines intersect is to use a technique called *substitution* — you *substitute* the y value from one equation for the y value in the other equation and then solve for x. Since you're looking for the place where x and y of each line are the same — that's where they intersect — then you can write the equation $y = y$, meaning that the y from the first line is equal to the y from the second line. Replace the y's with what they're equal to in each equation, and solve for the value of x that works.

Find the intersection of the lines $3x - y = 5$ and $x + y = -1$. (This is the same problem graphed in the preceding section.)

1. **Put each equation in the slope-intercept form, which is a way of solving each equation for y.**

   $3x - y = 5$ can be written as $y = 3x - 5$

   $x + y = -1$ can be written as $y = -x - 1$.

   The lines are not parallel, and their slopes are different, so there will be a point of intersection.

2. **Set the y points equal and solve.**

   $y = 3x - 5$ and $y = -x - 1$.

   Substitute what y is equal to in the first equation with the y in the second equation:

   $3x - 5 = -x - 1$.

3. **Solve for the value of x.**

   Add x to each side and add 5 to each side.

   $3x + x - 5 + 5 = -x + x - 1 + 5$.

   $4x = 4$

   $x = 1$

Substitute that 1 into either equation to find that $y = -2$. The lines intersect at the point $(1,-2)$.

This is how the solution can be found without even graphing. If the lines are parallel, it's apparent immediately because their slopes are the same. If that's the case, stop; there's no solution. Also, if the two equations are just two different ways of naming the same line, then this will be apparent, too. The equations will be exactly the same in the slope-intercept form.

Curling Up with Parabolas

Parabolas are nice, U-shaped curves. They are the graphs of quadratic expressions set equal to y. Picture a heavy cable hanging between towers of a suspension bridge; the gentle U-shaped curve is a parabola. The reflectors in headlights have parabola-like curves running through them. McDonald's golden arches are parabolas. The abundance of manufactured parabolas points to the fact that the properties responsible for creating a parabola often occur naturally. Mathematicians are able to put an equation to this natural phenomenon.

Parabolas have a highest point *or* a lowest point called the *vertex*. The curve is lower on the left and right of a vertex that is the highest point and it's higher to the left and right of a vertex that is the lowest point.

Trying out the basic parabola

My favorite name for the parabola with the equation $y = x^2$ is the *basic parabola*. Figure 19-16 shows a graph of this formula. This equation says that the y-coordinate of every point on the parabola is the square of the x-coordinate. Notice that whether x is positive or negative, the y is a square of it and is positive.

The vertex of the parabola in Figure 19-16 is at the origin, $(0,0)$, and it curves upward. You can make this parabola steeper or flatter by multiplying the x^2 by certain numbers. If you multiply by numbers bigger than one, it makes the parabola steeper. If you multiply by numbers between zero and one (proper fractions), then it makes the parabola flatter as Figure 19-17 shows. Making it steeper or flatter than the basic parabola helps the parabola fit different applications. The flatter ones are more like the curve of a headlight reflector. The steeper ones could be models for the time it takes to swim a certain distance, depending on your age.

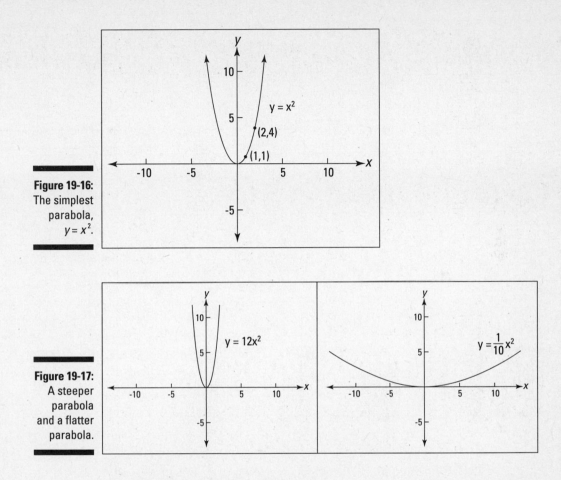

Figure 19-16:
The simplest parabola, $y = x^2$.

Figure 19-17:
A steeper parabola and a flatter parabola.

You can make the parabola open downward by multiplying the x^2 by a negative number, and make it steeper or flatter than the basic parabola — in a downward direction.

Putting the vertex on an axis

The basic parabola, $y = x^2$, can be slid around: left, right, up, down, placing the vertex somewhere else on an axis and not changing the general shape.

If you change the basic equation by adding a number to the x^2, such as: $y = x^2 + 3$, $y = x^2 + 8$, $y = x^2 - 5$, $y = x^2 - 1$, then the parabola moves up and down the y-axis. Note that adding a negative number is also part of this. These manipulations help make a parabola fit the model of a certain situation. Not everything starts at zero. Figure 19-18 shows several parabolas, none of which starts at zero.

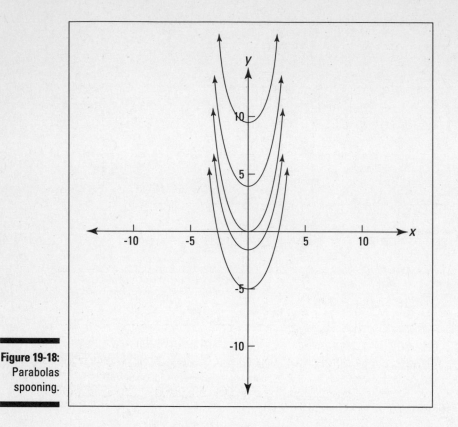

Figure 19-18:
Parabolas
spooning.

If you change the equation by adding a number to the x first and *then* squaring the expression, such as $y = (x + 3)^2$, $y = (x + 8)^2$, $y = (x - 5)^2$, or $y = (x - 1)^2$, you move the graph to the left or right of where the basic parabola lies. Using a +3 as in the equation $y = (x + 3)^2$ moves the graph to the *left,* and using a –3 as in the equation $y = (x - 3)^2$ moves the graph to the *right*. It's the opposite of what you might expect, but it works this way consistently. See Figure 19-19 just ahead.

Sliding and multiplying

You can combine the two operations of changing the steepness of a parabola and moving the vertex. These change the basic parabola to suit your purposes.

The equation used to model a situation may require a steep parabola because the changes happen rapidly, and it may require that the starting point be at eight feet, not zero feet. By moving the parabola around and changing the shape, you can get a better fit for the info you want to demonstrate.

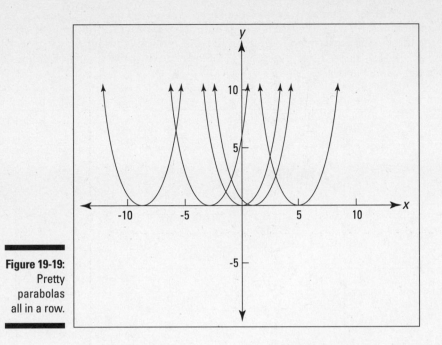

Figure 19-19:
Pretty
parabolas
all in a row.

The following equations and their graphs are shown in Figure 19-20.

$y = 3x^2 - 2$: The 3 multiplying the x^2 makes the parabola steeper and the -2 moves the vertex down to $(0,-2)$

$y = \frac{1}{4}x^2 + 1$: Makes it flatter and moves the vertex up to $(0,1)$

$y = -5x^2 + 3$: Makes it steeper, go downward, and moves the vertex to $(0,3)$

$y = 2(x-1)^2$: Makes it steeper and moves the vertex right to $(1,0)$

$y = -\frac{1}{3}(x+4)^2$: Makes it flatter, goes downward, and moves the vertex left to $(-4,0)$

$y = -\frac{1}{20}x^2 + 5$: Makes it flatter, goes downward, and moves the vertex to $(0,5)$

Generating the general form of a parabola

The most general equation of a parabola, which takes into account all the things that can be done to it, is

$$y = a(x-h)^2 + k$$

This is a standard way to write the equation of a parabola so that it gives you all the information you need to graph it. If the equation is in this form, then the values of *a*, *h* and *k* tell the story. They indicate how the parabola differs from the basic parabola.

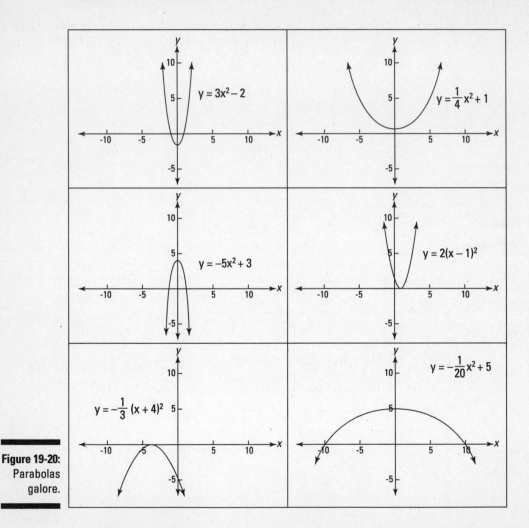

Figure 19-20:
Parabolas
galore.

✔ The *a* factor makes the parabola steeper or flatter, and it makes the parabola open upward when positive and downward when negative.

✔ The *h* tells how far left or right the vertex moves (+*h* goes left; −*h* goes right).

✔ The *k* tells how far up or down the vertex moves.

The following examples show how all this fits together. Start with the values of *h* and *k,* which give you the coordinates of the vertex. Then consider the steepness and flatness and whether it opens up or down.

✔ $y = (x - 3)^2 + 2$

The vertex of this parabola is 3 units to the right of the origin and 2 units up, so it's at (3, 2). The parabola opens upward, as shown in Figure 19-21.

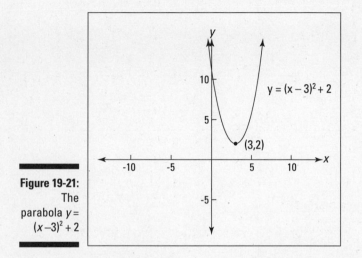

Figure 19-21:
The parabola $y = (x-3)^2 + 2$

✔ $y = \frac{-1}{4}(x + 1)^2 + 3$

The vertex of this parabola is 1 unit to the left of the origin and 3 units up. The parabola opens downward and is "flatter," as Figure 19-22 shows.

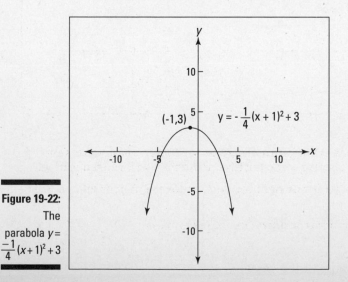

Figure 19-22:
The parabola $y = \frac{-1}{4}(x+1)^2 + 3$

Part V
The Part of Tens

The 5th Wave   By Rich Tennant

"GET READY, I THINK THEY'RE STARTING TO DRIFT."

In this part . . .

The Ten Most Common Errors, Ten Ways to Factor a Quadratic Equation, Ten Rules of Divisibility, Ten Steps to Solving a Story Problem, the Ten Commandments. Oops! How did that last one get in there?

The Part of Tens offers some quick, concise references to help you get organized and proceed correctly.

Chapter 20

The Ten Most Common Errors

So much algebra is done in the world: Just about everyone who advances beyond elementary school takes an algebra class, so the sheer number of people who use algebra means that a large number of errors are made. Forgetting some of the more obscure rules or confusing one rule with another is easy to do. But some errors occur because the error seems to be an easier way to do the problem. Not right, but easier — the path of least resistance. This usually happens when a rule isn't the same as your natural inclination. Most algebra rules seem to make sense, so they aren't hard to remember. Some, though, go against the grain.

The main errors occur while working expanding-type operations: distributing, or squaring binomials, or breaking up fractions, or raising to powers. The other big error area is in dealing with negatives. Watch out for those negative vibes.

Missing Middle Term

A squared binomial has three terms in the answer. The term that gets left out is the middle term: the part you get when multiplying the two *outer* terms together and the two *inner* terms together and finding their sum. Often, just the first and last separate terms are squared, and the middle term is just forgotten.

Right: $(a + b)^2 = a^2 + 2ab + b^2$

Wrong: $(a + b)^2 \neq a^2 + b^2$

Go to Chapter 8 for more information on squaring binomials.

Distributing

Distributing a number or a negative sign over two or more terms in a parenthesis can cause problems if you forget to distribute the value over *every single term* in the parenthesis. The errors come in when you stop multiplying the terms in the parenthesis before you get to the end.

Right: $x - 2(y + z - w) = x - 2y - 2z + 2w$

Wrong: $x - 2(y + z - w) \neq x - 2y + z - w$

There's more on distributing in Chapter 8.

Breaking Up Fractions

Splitting a fraction into several smaller pieces is all right as long as each piece has a term from the numerator (top) and the entire bottom (denominator). You can't split up the denominator.

Right: $\dfrac{x + y}{a + b} = \dfrac{x}{a + b} + \dfrac{y}{a + b}$

Wrong: $\dfrac{x + y}{a + b} \neq \dfrac{x}{a} + \dfrac{y}{b}$

Go to Chapter 3 for more on dealing with fractions.

Breaking Up Radicals

If the expression under a radical has values multiplied together or divided, then the radical can be split up into radicals that multiply or divide. You can't split up addition or subtraction, however, under a radical.

Right: $\sqrt{a^2 + b^2} = \sqrt{a^2 + b^2}$

Wrong: $\sqrt{a^2 + b^2} \neq \sqrt{a^2} + \sqrt{b^2}$

For more on radicals, go to Chapter 4.

Order of Operations

The order of operations instructs you to raise the expression to a power *before* you add or subtract. A negative in front of a term is in the same category as subtracting. It has to be done last. If you want the negative raised to the power, too, then include it in parentheses with the rest of the value.

Right: $-3^2 = -9$

Right: $(-3)^2 = 9$

Wrong: $-3^2 \neq 9$

The order of operations is discussed fully in Chapter 5.

Fractional Exponents

A fractional exponent has the power on the top of the fraction and the root on the bottom. Remember, when writing $\sqrt{x}$ as a term with a fractional exponent, $\sqrt{x} = x^{1/2}$. A fractional exponent indicates that there's a radical involved. The two in the fractional exponent is on the bottom — the root always is the bottom number.

Right: $\sqrt[5]{x^3} = x^{3/5}$

Wrong: $\sqrt[5]{x^3} \neq x^{5/3}$

Check out Chapter 4 for more on fractional exponents.

Multiplying Bases Together

When multiplying numbers with exponents and the same base, you add the *exponents* and leave the base as it is. The bases never get multiplied together.

Right: $2^3 \cdot 2^4 = 2^7$

Wrong: $2^3 \cdot 2^4 \neq 4^7$

Go to Chapter 4 for more on multiplying numbers with exponents and the same base.

A Power to a Power

To raise a value that has a power to another power, *multiply* the exponents to raise the whole term to another power. Don't raise the exponent itself to a power — it's the base that's being raised, not the exponent.

> **Right:** $(x^2)^4 = x^8$
>
> **Wrong:** $(x^2)^4 \neq x^{16}$

Chapter 4 is the place to go for more on powers.

Reducing

When reducing fractions with a numerator that has more than one term separated by addition or subtraction, whatever you're reducing the fraction by has to divide *every single term* evenly in both the numerator and denominator.

> **Right:** $\dfrac{(4 + 6x)}{4} = \dfrac{(2 + 3x)}{2}$
>
> **Wrong:** $\dfrac{(4 + 6x)}{4} \neq \dfrac{(2 + 6x)}{2}$

Go to Chapter 3 if you want more information on fractions.

Negative Exponents

When changing fractions to expressions with negative exponents, give every single factor in the denominator a negative exponent.

> **Right:** $\dfrac{1}{2ab^2} = 2^{-1}a^{-1}b^{-2}$
>
> **Wrong:** $\dfrac{1}{2ab^2} \neq 2a^{-1}b^{-2}$

There's more on negative exponents in Chapter 4.

Chapter 21

Ten Ways to Factor a Polynomial

Polynomials can be factored in many ways, depending on the number of terms and how they're made up. But it's usually easier if you have a checklist to refer to while you're doing the factoring. If a polynomial has two terms (which means it's a binomial), there are only four possible ways to factor it. If none of those methods works, then the expression can't be factored. The techniques in this chapter are those most commonly used and should get you through most factoring situations.

Two Terms with a GCF

In an expression with two terms, look for some*thing* or some*things* that each term has in common. Then factor that greatest common factor out of each term and write the result as a product of the GCF and the division thus:

$$ab + ac = a(b + c)$$

The Difference of Two Squares

When an expression is comprised of two terms that are both perfect squares separated by subtraction, they factor into the sum and difference of the square roots of the terms:

$$a^2 - b^2 = (a + b)(a - b)$$

The Difference of Two Cubes

When you're dealing with perfect cubes separated by subtraction, you can factor the two terms into the product of a binomial and a trinomial with the pattern:

$$a^3 - b^3 = (a - b)(a^2 + ab + b^2)$$

The Sum of Two Cubes

Expressions composed of two terms that are perfect cubes separated by addition factor into the product of a binomial and a trinomial, following this pattern:

$$a^3 + b^3 = (a + b)(a^2 - ab + b^2)$$

Three Terms with a GCF

When you encounter three terms in an expression, look for something or somethings that each term has in common with the others. You can then factor that greatest common factor out of each term and write the result as a product of the GCF and the factored terms.

$$ab + ac + ad = a(b + c + d)$$

Three Terms with unFOIL

When three terms in a quadratic expression are of the form $ax^2 + bx + c$, look for a way to factor the terms into the product of two binomials using unFOIL (check out Chapters 10 and 14 for a refresher on unFOILing).

$$ax^2 + bx + c = (dx + e)(fx + g)$$

The x is the variable. The a, b, c are real numbers that are not zero. The d, e, f, g represent real numbers in the factorization. The $a = d \cdot f$ and the $c = e \cdot g$.

Changing to Quadratic-Like

An expression with three terms in the form $ax^{2n} + bx^n + c$ dictates that you first write the expression as a comparable quadratic (second-degree) expression such as $ax^2 + bx + c$, and then look for a way to factor it into the product of two binomials using unFOIL.

The a, b, c are all real numbers, not equal to zero. The n is a counting number, 1, 2, 3, and so on, so the first exponent is always a counting number that's twice the second exponent. Go to Chapters 11 and 14 for more on this method.

Four or More Terms with a GCF

When you encounter expressions with four or more terms, looking for a greatest common factor (or factors) is the first thing to do. If you find a GCF, factor it out of each term and write the result as a product of the GCF and the factored terms, like so:

$$ab + ac + ad + ae = a(b + c + d + e)$$

Four or More Terms with Equal Grouping

When you run into expressions with four or six or eight or some even number of terms, try grouping the terms into equal-size groups, each with its own greatest common factor. Then look for a GCF in the newly formed terms. For example:

$$ax + ay + bx + by + cx + cy =$$
$$a(x + y) + b(x + y) + c(x + y) =$$
$$(x + y)(a + b + c)$$

Go to Chapter 11 for more details on grouping.

Four or More Terms with Unequal Grouping

When there are four or more terms in an expression and there aren't any ways to arrange them in equal groups that have a GCF, then look for unequal groupings that may factor into perfect squares or cubes.

Try to factor the new groups using the rules for factoring the difference of squares or the difference of the sum of cubes, found earlier in this chapter and in Chapter 11.

$$x^2 - a^2 + 2ab - b^2 = x^2 - (a^2 - 2ab + b^2) = x^2 - (a - b)^2 =$$
$$[x - (a - b)][x + (a - b)]$$

Chapter 22

Ten Divisibility Rules

Divisibility rules are shortcuts to seeing whether a number divides evenly by another number. This is really helpful when you're reducing fractions or determining whether everyone in a group gets the same number of pieces of candy. A number that is evenly divisible by *all* the numbers with rules in this chapter is 3,960. In fact, it's the smallest number divisible by all of them.

When I use the term *divisible,* I mean *evenly* divisible, with no remainder.

Divisibility by 2

A number is divisible by 2 if the last digit in the number is 0, 2, 4, 6 or 8:

- ✔ 1,234 is divisible by 2 because it ends in 4.
- ✔ 2,345 is not divisible by 2 because it ends in 5.
- ✔ 3,960 is divisible by 2 because it ends in 0.

Divisibility by 3

A number is divisible by 3 if the sum of the digits in the number is divisible by 3:

- ✔ 1,234 is not divisible by 3 because $1 + 2 + 3 + 4 = 10$.
- ✔ 3,345 is divisible by 3 because the sum of the digits in the number is 15.
- ✔ 120,000,000,000,000 is divisible by 3 because the sum of the digits is 3.
- ✔ 3,960 is divisible by 3 because the sum of the digits is 18.

Divisibility by 4

A number is divisible by 4 if the last two digits in the number form a number divisible by 4:

- ✔ 1,234 is not divisible by 4 because 34 is not divisible by 4.
- ✔ 121,212 is divisible by 4 because 12 is divisible by 4.
- ✔ 444,444,444,444,414 is not divisible by 4 because 14 is not divisible by 4.
- ✔ 3,960 is divisible by 4 because 60 is divisible by 4.

Divisibility by 5

A number is divisible by 5 if the last digit is 0 or 5:

- ✔ 12,345 is divisible by 5 because the last digit is 5.
- ✔ 555,552 is not divisible by 5 because the last digit is 2.
- ✔ 342,000,460 is divisible by 5 because the last digit is 0.
- ✔ 3,960 is divisible by 5 because the last digit is 0.

Divisibility by 6

A number is divisible by 6 if it is divisible by both 2 and 3. That means it has to end in 0, 2, 4, 6, or 8 and the sum of the digits has to be divisible by 3:

- ✔ 3,000,000 is divisible by 6 because it ends in 0 and the sum of the digits is 3.
- ✔ 111,111,102 is divisible by 6 because it ends in 2 and the sum of the digits is 9.
- ✔ 666,666,010 is not divisible by 6 because the sum of the digits is 37.
- ✔ 3,960 is divisible by 6 because it ends in 0 and the sum of the digits is 18.

Divisibility by 8

A number is divisible by 8 if the last three digits form a number divisible by 8:

- ✔ 5,005,808 is divisible by 8 because 808 is divisible by 8.
- ✔ 888,888,111 is not divisible by 8 because 111 is not divisible by 8.
- ✔ 3,960 is divisible by 8 because 960 is divisible by 8.

Divisibility by 9

A number is divisible by 9 if the sum of the digits of the number is divisible by 9:

- ✔ 123,111 is divisible by 9 because the sum of the digits is 9.
- ✔ 111,111 is not divisible by 9 because the sum of the digits is 6.
- ✔ 108,000 is divisible by 9 because the sum of the digits is 9.
- ✔ 3,960 is divisible by 9 because the sum of the digits is 18.

Divisibility by 10

A number is divisible by 10 if it ends in 0:

- ✔ 111,110 is divisible by 10.
- ✔ 100,000,003 is not divisible by 10.
- ✔ 3,960 is divisible by 10 because it ends in 0.

Divisibility by 11

A number is divisible by 11 if the sums of the alternate digits are different by 0, 11, 22, or 33, or any two-digit multiple of 11. In other words, say you have a six-digit number: Add up the first, third, fifth digits — the odd ones. Then add the digits in the even places — second, fourth, sixth. Then subtract those totals from each other, and if the answer is a multiple of 11, the original number is divisible by 11. I'm not saying that this is a tip you'll use everyday, I'm just telling you that it works.

✔ 146,322 is divisible by 11 because the alternate digits 1, 6, 2 add up to 9 and the alternate digits 4, 3, 2 add up to 9. The difference between 9 and 9 is 0.

✔ 818,290 is divisible by 11 because the alternate digits add up to 25 and 3, and the difference between the two sums is 22.

✔ 3,960 is divisible by 11 because the alternate digits add up to 9 in both cases, so the difference is 0.

Divisibility by 12

A number is divisible by 12 if the last two digits form a number divisible by 4 and if the sum of the digits is divisible by 3:

✔ 333,216 is divisible by 12 because 16 is divisible by 4 and the sum of the digits is 18, which is divisible by 3.

✔ 2,000,010,000 is divisible by 12 because 00 is divisible by 4 and the sum of the digits is 3.

✔ 3,960 is divisible by 12 because 60 is divisible by 4 and the sum of the digits is 18.

Chapter 23

Ten Tips for Dealing with Story Problems

*E*ach story problem is unique unto itself. Just as my fingerprints are different than yours, so are story problems different from one another. But as unique as fingerprints and story problems are, they share similarities and patterns. Making the most of these similarities and patterns makes for success in dealing with story problems. The list of suggestions for dealing with story problems helps you discover these similarities. Use as many or as few of these as you need to work through the story problems you encounter.

Draw a Picture

Many people respond well to visual prompts. Draw a picture — nothing fancy — to represent what's going on in a story. Label your picture with numbers or names or other information that helps you make sense of the situation. Fill it in more or change the drawing as you set up an equation for the problem.

Make a List

Try out some answers. If the problem asks how many of the 100 people were children, make a list trying some combinations: 90 + 10, 80 + 20, and so on. You can sometimes stumble on the solution through your list-making, and you can then work backwards to set up a systematic way of solving the equation. Even if you don't find the answer this way, it will still give you a sense of what the answer might be. You can figure out what's in the ball park.

Assign Variables to Represent Numbers

Let variables (letters) represent numbers. A variable can represent the length of a boat or the number of people, but it can't represent the boat itself or a person. You can choose the letters so they can help make sense of the problem. For example, you can let k represent Ken's height; just don't let it represent Ken.

Translate Conjunctions and Verbs

You can use clues from a problem to help you set up an equation by translating words into math symbols. Generally, you can let a plus sign replace words such as *and* or *increased by,* a minus sign replaces *less than* or *taken away.* Use $2x$ for *twice,* and the equal sign for *are, has, is,* and other verbs.

You can often write an equation in the same order as the clue words in the problem. For example:

Stephan has six more than twice as many books as Alicia.

You can write this as $s = 6 + 2a$ if you let s represent the number of books that Stephan has and a represent the number of books that Alicia has. See Chapter 18 for more on words and how they translate into algebra.

Look at the Last Sentence

Look at the last sentence of the problem. It usually tells you what the variables should stand for. The last sentence may also let you know if you should use a traditional formula for area, distance, interest, or volume. For example:

> Marilee and Scott ran in a race. Marilee finished two minutes before Scott, but she ran one less kilometer than Scott did. If they ran at the same rate and the total distance they ran (added together) was nine kilometers, then how long did it take them?

Just look at all those words. Go to the last sentence — and even the last phrase of the last sentence. It tells you that you're looking for the amount of time it took. The formula the last sentence suggests is $d=rt$ — distance equals rate times time.

Find a Formula

When possible, use a formula as your equation or as part of your equation. Formulas are a good place to start to set up relationships. Keep the standard formulas in a place where you can use them for quick reference. (You can use the handy, tear-out Cheat Sheet at the front of this book for that purpose.) Be familiar with what the variables in the formula stand for.

Simplify by Substituting

Look for variables with a relationship to another variable and try to express one of those variables in terms of the other. For example, if one side of a triangle is twice as long as another, you can express those two sides as x and $2x$ rather than x and y.

Solve an Equation

Translate the story into an equation that represents the situation and relationships expressed in the problem. Solve the equation carefully, using the rules of algebra.

Check for Sense

When you get an answer, decide whether it makes sense within the context of the problem. If you're solving for the height of a man, and your answer comes out to be 40 feet, you probably made an error somewhere. Having an answer make sense doesn't guarantee that it's a correct answer, but it's the first check to tell if it *isn't* correct.

Check for Accuracy

If you determine that your answer makes sense, check the algebra. Do that by putting the solution back into the original equation and checking. If that works, then work your answer through the written story problem to see if it works out with all of the situations and relationships.

Glossary

absolute value: An operation that tells you how far a number is from zero.

additive inverse: Number with the same numerical part but the opposite sign (plus or minus) of the given number. If zero is the sum of two numbers, then these two numbers are *additive inverses* of one another.

area: Measure of a specified region in a plane.

associative property: Characteristic of addition and multiplication that allows the grouping of terms to change without affecting the result.

base: Value multiplied repeatedly in an exponential expression.

binary operation: Process requiring two values to produce a third value.

binomial: Two terms separated by addition or subtraction.

circumference: Distance around the outside of a circle.

coefficient: Number multiplied by a variable.

combinations: Method of counting that tells how many ways a designated number of objects can be selected from a given set.

common denominator: Same value on the bottom of more than one fraction.

commutative property: Characteristic of addition and multiplication that allows the order of the values in an operation to be changed without affecting the result.

composite number: Whole number larger than one that isn't prime.

constant: Variable or number that never changes in value in an expression.

coordinate: Part of an ordered pair that designates a point's location on a coordinate plane.

coordinate plane: Flat region determined by two intersecting, perpendicular, numbered lines called *axes*.

cube: 1 Third power of a number. Result of multiplying a number by itself three times. **2** A three-dimensional solid with six square surfaces.

cube root: Number that you can multiply by itself three times to get a given number. For example, the cube root of eight is two because two multiplied by itself three times equals eight.

cubic: Adjective that describes an expression in which the highest power is three.

decimal: 1 Fraction with an unwritten denominator of 10 indicated by the decimal point. **2** Adjective describing the number system that is organized in increments of ten.

degree of an angle: Number of degrees between 1 and 360 that comprise the measure of an angle.

degree of an expression: Highest power occurring in the expression.

denominator: Bottom number of a fraction.

diameter: Longest distance across a circle.

difference: Result of subtraction.

digit: Numerals from zero through nine, so called because they were originally counted on the fingers.

distributive property: Characteristic of multiplication and addition that allows for the multiplication of each individual term in a grouped series by a term outside of the grouping without changing the value of the expression.

divisible: One number can be divided by another with no remainder.

double root: Solution that appears twice when solving an equation because the related factor appears twice in the factored form. For example, in $(y - 2)(y - 2) = 0$, $y = 2$ is a double root or solution. In $(y - 2)(y - 2)(x + 3) = 0$, $y = 2$ is still a double root or solution.

equation: Mathematical statement with an equal sign showing that two values are equal.

equivalent fractions: Fractions equal to one another, even though they may have different denominators.

exponent: Also known as *power*, value in smaller type found above and to the right of the base that indicates repeated multiplication of the base.

expression: Combination of values (variables, numbers, and/or constants) and operation(s).

factor: 1 One of the values in a multiplication operation. **2** To rewrite an algebraic expression as a product.

factorial: Operation that multiplies a whole number by every counting number smaller than it.

formula: Rule or method that is accepted as true and used over and over in common applications.

fraction: Quantity expressed as a numerator, the value on top of the bar, and a denominator, the value below the bar that determines how many make one.

graph: Plotted figure in a plane.

greatest common factor: (GCF) Largest possible value that evenly divides each term of an expression containing two or more terms.

grouping symbol: Parentheses, brackets, and braces that can affect the order of operations. Terms and operations within the grouping symbol take precedence.

hypotenuse: Longest side of right triangle.

inequality: Relationship between two unequal values.

infinite: Without end; uncountable.

integer: A positive or negative whole number or zero; numbers starting with zero and going up or down in increments of one.

intercept: Point where a graph crosses the x-axis or y-axis.

intersection: Point shared by two lines.

irrational number: Number with no fractional equivalent whose decimal never repeats or terminates.

line: All the points in the coordinate plane that satisfy a linear equation.

linear: Adjective describing expression or equation in which the highest power of any variable is one. Constants can be higher powers. For example, $x + y = 4$ is linear.

lowest terms: Status of a fraction whose numerator and denominator cannot be divided evenly by the same number. A fraction with a numerator of 12 and a denominator of 25 is in lowest terms.

mixed number: Improper fraction written as a whole number alongside a fraction.

monomial: An expression with only one term.

multiple: A number evenly divisible by a specific factor. For example, the numbers 14 and 21 are multiples of 7.

multiplication property of zero: (MPZ) Rule stating if the product of two numbers is zero, then one of the numbers must be zero.

multiplicative inverse: Also known as a reciprocal, one of two numbers whose product is one. The *reciprocal* of a number is that particular number in the denominator of a fraction with a value of one in the numerator.

natural numbers: Also called *counting* numbers, values starting with one and increasing by one.

negative number: Any quantity that is less than zero; usually preceded by a minus sign.

negative reciprocals: Two numbers, one positive and one negative, whose product is negative one.

numerator: The top number in a fraction.

operation: Mathematical process, such as addition, subtraction, multiplication, and division, performed on one or more quantities.

opposite: 1 The opposite of an *operation* is another operation that gets you back where you started. **2** The opposite of a *number* is the additive inverse — the same absolute value of a number with a different sign.

ordered pair: Two values inside parentheses and separated by a comma that indicate the position of a point in the coordinate plane.

origin: Point of intersection of the *x*-axis and *y*-axis in a coordinate plane.

parallel lines: Lines that never intersect and are always the same distance apart.

percent: Fractions with a denominator of 100. The percentage is the numerator of the fraction — how many out of 100.

perimeter: Total distance around the outside of a region or area.

permutation: Counting method that determines the number of ordered arrangements there are when a certain number of objects are selected from a given set.

perpendicular lines: Lines that form a 90-degree angle at their intersection.

pi (π): a letter from the Greek alphabet, that refers to the relationship between the diameter and the circumference of a circle. It is approximately 3.14 or $^{22}/_{7}$.

polynomial: Expression with one or more terms.

positive number: Any quantity greater than zero.

power: Value of an exponent indicating the number of times the base is multiplied by itself.

prime factorization: Process of finding the prime numbers that, when multiplied together, produce a given composite number.

prime number: Whole number larger than the number one that can be divided evenly only by itself and one.

principal square root: Positive number that when multiplied by itself produces a given positive number. For example, the square roots of 25 are 5 and –5, but the *principal* square root of 25 is only 5.

product: Result of multiplication.

proper fraction: Fraction whose value is less than one. The numerator is always smaller than the denominator.

proportion: Equation showing that two ratios are equal to one another, such as one is to two as three is to six $\left(1{:}2 = 3{:}6 \text{ or } \frac{1}{2} = \frac{3}{6}\right)$.

Pythagorean theorem: Formula specific to right triangles stating that the hypotenuse (c) squared is equal to the sum of the squares of the remaining sides (*a* and *b*): $a^2 + b^2 = c^2$.

quadrant: One of four regions in a coordinate plane defined by the *x*-axis and the *y*-axis.

quadratic: Also known as *second degree*, expression or equation in which the highest power is two. The degree is two.

quotient: Result of division.

radical ($\sqrt{}$): Symbol for the operation to find a square root.

radius: Distance from the center of a circle to its outer edge; half its diameter.

rational number: Quantity, positive or negative, that can be written as a fraction; its decimal equivalent terminates or repeats.

reciprocal: Multiplicative inverse of a given number. The product of a given number and its reciprocal is always one.

rectangle: Four-sided plane figure with all right angles; its opposite sides are equal to one another in length.

reduce: Process in which a common factor of the numerator and denominator of a fraction is divided out, leaving an equivalent fraction.

relatively prime: Two numbers that have no factors in common other than the number one.

remainder: Value that is left over when one number is divided by another.

right angle: 90-degree angle.

right triangle: Three-sided plane figure with a 90-degree angle (right angle).

root: Value that multiplied by itself a number of times results in the value or number wanted, such as two is the *root* of four because two multiplied by itself produces four.

rounding: Approximating value to the nearest digit or decimal place, such as rounding 14.9 up to 15.

scientific notation: A standard way of writing very large and very small numbers as the product of two values — a number between one and ten and a power of ten.

semiperimeter: Half the perimeter.

sign: Symbol indicating whether a value is positive (+) or negative (–).

simplify: To combine all that can be combined and put an expression in its most easily understandable form.

slope: Number indicating the measure of a line's steepness or slant and whether it rises or falls.

solution of equation: Value(s) of the variable which make the equation a true statement.

solve: Find the answer or what number the variable stands for.

square: **1** Four-sided figure in a plane. All the sides of a square are the same length and meet at right angles. **2** Value with an exponent of two. **3** (perfect square) Product of another number times itself.

square root: Value resulting in a given value when multiplied by itself. For example, the square root of four is two because two times two is four, which is a perfect square.

substitution: Method of replacing a value with its equivalent.

sum: Result of addition.

symmetric property: Characteristic of equations that allows for the exchange of the value(s) on one side of the equal sign with the value(s) on the other side (quantities on the right go to the left; quantities on the left go to the right) without changing the truth of the equation: If $x = y$, then $y = x$.

synthetic division: Short-cut division process in which only the coefficients of the terms in an expression are used. The answer is obtained by multiplying and adding.

term: Constants, numbers, and/or variables connected to one another by multiplication or division.

trinomial: Expression with three terms. Each term is separated from the others by addition or subtraction.

value: A numeric equivalence or worth of an expression or variable.

variable: Letter representing an unknown number or what you're solving for in an algebra problem.

vertex: Corner of a figure; where two sides intersect to form an angle.

volume: Measurement of the amount of space within a three-dimensional solid figure.

whole numbers: All natural numbers plus zero. Start with zero and add one repeatedly to find the whole number spectrum (0, 1, 2, 3 . . .).

Index

FOR DUMMIES®

A world of resources to help you grow

HOME, GARDEN & HOBBIES

Feng Shui
0-7645-5295-3

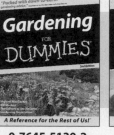

Gardening
0-7645-5130-2

Guitar
0-7645-5106-X

Also available:

Auto Repair For Dummies
(0-7645-5089-6)

Chess For Dummies
(0-7645-5003-9)

Home Maintenance For
Dummies
(0-7645-5215-5)

Organizing For Dummies
(0-7645-5300-3)

Piano For Dummies
(0-7645-5105-1)

Poker For Dummies
(0-7645-5232-5)

Quilting For Dummies
(0-7645-5118-3)

Rock Guitar For Dummies
(0-7645-5356-9)

Roses For Dummies
(0-7645-5202-3)

Sewing For Dummies
(0-7645-5137-X)

FOOD & WINE

Cooking
0-7645-5250-3

Cookies
0-7645-5390-9

Wine
0-7645-5114-0

Also available:

Bartending For Dummies
(0-7645-5051-9)

Chinese Cooking For
Dummies
(0-7645-5247-3)

Christmas Cooking For
Dummies
(0-7645-5407-7)

Diabetes Cookbook For
Dummies
(0-7645-5230-9)

Grilling For Dummies
(0-7645-5076-4)

Low-Fat Cooking For
Dummies
(0-7645-5035-7)

Slow Cookers For Dummies
(0-7645-5240-6)

TRAVEL

Italy
0-7645-5453-0

Hawaii
0-7645-5438-7

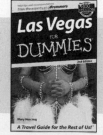

Las Vegas
0-7645-5448-4

Also available:

America's National Parks For
Dummies
(0-7645-6204-5)

Caribbean For Dummies
(0-7645-5445-X)

Cruise Vacations For
Dummies 2003
(0-7645-5459-X)

Europe For Dummies
(0-7645-5456-5)

Ireland For Dummies
(0-7645-6199-5)

France For Dummies
(0-7645-6292-4)

London For Dummies
(0-7645-5416-6)

Mexico's Beach Resorts For
Dummies
(0-7645-6262-2)

Paris For Dummies
(0-7645-5494-8)

RV Vacations For Dummies
(0-7645-5443-3)

Walt Disney World & Orlando
For Dummies
(0-7645-5444-1)

Available wherever books are sold. Go to www.dummies.com or call 1-877-762-2974 to order direct.

FOR DUMMIES®

Plain-English solutions for everyday challenges

COMPUTER BASICS

0-7645-0838-5

0-7645-1663-9

0-7645-1548-9

Also available:

PCs All-in-One Desk Reference For Dummies (0-7645-0791-5)

Pocket PC For Dummies (0-7645-1640-X)

Treo and Visor For Dummies (0-7645-1673-6)

Troubleshooting Your PC For Dummies (0-7645-1669-8)

Upgrading & Fixing PCs For Dummies (0-7645-1665-5)

Windows XP For Dummies (0-7645-0893-8)

Windows XP For Dummies Quick Reference (0-7645-0897-0)

BUSINESS SOFTWARE

0-7645-0822-9

0-7645-0839-3

0-7645-0819-9

Also available:

Excel Data Analysis For Dummies (0-7645-1661-2)

Excel 2002 All-in-One Desk Reference For Dummies (0-7645-1794-5)

Excel 2002 For Dummies Quick Reference (0-7645-0829-6)

GoldMine "X" For Dummies (0-7645-0845-8)

Microsoft CRM For Dummies (0-7645-1698-1)

Microsoft Project 2002 For Dummies (0-7645-1628-0)

Office XP For Dummies (0-7645-0830-X)

Outlook 2002 For Dummies (0-7645-0828-8)

Get smart! Visit www.dummies.com

- **Find listings of even more *For Dummies* titles**

- **Browse online articles**

- **Sign up for Dummies eTips™**

- **Check out *For Dummies* fitness videos and other products**

- **Order from our online bookstore**

Available wherever books are sold. Go to www.dummies.com or call 1-877-762-2974 to order direct.

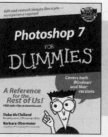